Advance Praise for *When AI Rules the World*

"In *When AI Rules the World*, Handel Jones sharpens the growing divergence between the US's and China's AI strategy. Through a comprehensive comparison of both countries, he clearly shows how AI is and will continue to have a disruptive impact on many sectors including military, industry, healthcare, AR/VR and connectivity. As a whole, the book paints a clarifying picture of how AI is acting as catalyst for a growing geo-political divide. A divide where China might not be leading yet on all accounts, but is progressing at a much faster pace, driven by huge amounts of data and large-scale investments in R&D. By putting innovations from all these sectors together, the book nicely outlines how we are in the middle of a geo-political race for AI technology, similar to the space race in the 1960s and beyond. However, this time AI is the rocket and data is the rocket fuel. The book's call for action is crucial. The disruptive nature of AI calls for national AI strategies, comprising large investments in R&D from the software level up to semiconductors and components, a vision I very much share."

—Dr. Luc Van den hove,
President and CEO, imec

"The world continues to be transformed by new software operating ever more sophisticated computers. But we have reached an inflection point in the technology's impact with Artificial Intelligence software that increasingly impacts computers' ability to emulate some key aspects of human intelligence. This technology has enormous potential impact in all fields of human activity. Mr. Handel Jones in his remarkable new book *When AI Rules the World* brilliantly describes the technology and its development but more importantly the global and societal implications for the future when advanced computers will largely replace humans in many activities. This book is not to be missed."

—Dr. Henry Kressel, Author, Technologist, Inventor, and long-term Senior Investing Partner at Warburg Pincus private equity firm

"AI is becoming a key part of our daily lives and we need to fully understand what the implications are. The book *When AI Rules The World* gives a deep understanding of the opportunities and threats of AI including the activities in China and is a must read book for professionals."

—Nicolas Dufourcq, CEO, Bpifrance

"This is the century of AI. At the end of this century all industries, including automotive, will be different. If you want to know the opportunities and threats of AI, if you want to understand how AI will rule the world and define the new political balance of the world, then read this book."

—Benedetto Vigna, CEO, Ferrari

"Despite the recent Russian aggression in the Ukraine, most forward geopolitical thinkers recognize that China possesses the greatest challenge to the United States this century. Mr. Handel Jones, a leading technologist and expert on China, raises the alarm on China's alarming lead in Artificial Intelligence technology. Jones's work, in simple terms, describes a world not far in the future, dominated by AI. He explores a dystopian picture of what Chinese dominance in this arena could mean, as well as a prescription on how the U.S. could retake the lead in these technologies through the combination of private industry and governmental leadership. *When AI Rules The World* is a must read for members of the U.S. Congress, investors, and Americans concerned about technological advancements in China that threaten the United States' role as the world's leading economic and military superpower."

—Stephen A. Kaplan, Retired Co-founder Oaktree Capital Management, Chairman of Nalpak Capital, Chairman DC United, and Co-owner of Swansea City Football Club

"Drawing upon his extensive experience in semiconductors, technology and China, Mr. Handel Jones creates a fact-supported narrative to make the case for artificial intelligence as the game changer for U.S. competition with China. Without a national initiative comparable to the post-Sputnik space race to the moon, he argues that the U.S. risks becoming an also-ran in advanced medical, automotive, military, education, and communication technologies. Mr. Jones provides a compelling argument for a broad U.S.-based pursuit of the AI-enabled technologies and applications."

—Walden C. Rhines, CEO Emeritus, Mentor, a Siemens Business; CEO, Cornami, Inc.

"AI is already transforming the world – it's moved from science fiction to something that is deeply reshaping all industries and the way we live and work. Handel Jones put together a compelling historical and comparative study of AI applications and policy in the US and China, explaining differences, successes and areas in need of a rethink. His extensive experience in both countries and technology enables him to outline a clear set of recommendations, contributing to the ongoing debate about how to drive technology leadership for the 21st century. *When AI Rules the World* is a must-read in 2022."

—Jean-Marc Chery, President & CEO, STMicroelectronics

"AI not only impacts our individual daily lives, but is also reshaping our global society. The book *When AI Rules the World* is an insightful read that provides thought provoking details on the potential impact that advancements in AI will have on our future society and culture."

—Dr. Anirudh Devgan, President and Chief Executive Officer, Cadence Design Systems, Inc.

WHEN AI RULES THE WORLD

CHINA, THE U.S., AND THE RACE TO CONTROL A SMART PLANET

WHEN AI RULES THE WORLD

CHINA, THE U.S., AND THE RACE TO CONTROL A SMART PLANET

HANDEL JONES

Published by Bombardier Books
An Imprint of Post Hill Press
ISBN: 978-1-64293-812-8
ISBN (eBook): 978-1-64293-813-5

When AI Rules the World:
China, the U.S., and the Race to Control a Smart Planet

Cover Design by Matt Margolis

Post Hill Press
New York • Nashville
posthillpress.com

Published in the United States of America
1 2 3 4 5 6 7 8 9 10

Contents

Introduction

On October 16th, 2021, the Financial Times broke a story with news that sent chills through the United States defense establishment: China had launched a hypersonic missile that could evade all American defense systems. The missile, which could fly more than 3,800 miles per hour and carry a nuclear weapon, circumnavigated the globe before speeding to its target.

Two things were particularly alarming about the news. First, a hypersonic vehicle could plow a suddenly shifting path, meaning that it could approach the United States from any number of directions. Missile defenses of the continental United States all point north and west over the Pacific, meaning they would fail to destroy missiles coming from the south. And because the Chinese hypersonic weapon could quickly change its path, even American defenses resituated to face south would fail to stop a Chinese attack. Second, and even more disturbing, was that the United States had been working on hypersonic technologies for decades, but American hypersonic development programs had somehow fallen years behind the Chinese.

If this were a one-off failure of American technology to surpass or even keep up with Chinese technology developments, it might be just one of many concerns facing the United States.

But it is more than that. Rapidly and systematically, China has broken new ground in technological achievements in industries ranging from defense to automotive vehicles to telecommunications to robotics to healthcare. With strong government leadership and government investments, China has placed itself in the once unimaginable position of potentially leading the United States in industry after industry. In particular, China is speeding ahead in artificial intelligence, the master technology that in the near future will make other technologies possible. If nothing else, I thought, an examination of AI development in China and the United States would provide a glimpse, for better or worse, into the future. So went the thinking that led me to write this book.

AI is going to have a far greater impact on the way people live than most Americans realize. By 2040, AI will impact every nation and business around the globe with dramatic and unexpected consequences. AI-empowered machines will have the ability to outperform the human brain just as excavating machines once replaced human muscles in the moving of earth. Eventually, people will be supported by virtual digital twins that compile and analyze data just like the human brain. As a result, digital twins will drastically increase people's memories and give them the power to complete mental tasks that they could not on their own. Autonomous vehicles, aided by AI, will be a mainstay of personal transportation in the near future. AI will also aid medical providers in monitoring, diagnosing, and treating diseases; improving nutrition; and extending lives. The resulting increases in longevity will, in turn, alter demographic trends and present society with a host of new quandaries.

Virtual reality, also powered by AI, will allow people to complete tasks they could not have done in the past, such as architects who can walk through and inspect a building that does not yet exist. Eventually, virtual reality will allow people to spend their lives in simulated worlds where material things that are important to us today will be available in virtual form, reducing the need to own many physical objects. Meanwhile, AI will bring about major changes in military capabilities, strategies, and threats. The armed forces of nations will be able to defeat or intimidate those of other nations because they can disable and defeat traditional military hardware. They will force their will upon nations whose militaries have fallen technologically behind. Most strikingly, these technical and social changes will not take place over 200 years, as did the last industrial revolution. They will take place over the next ten to twenty years.

I have been in a unique position to observe AI development and implementation in China and the United States in recent years. A veteran of four decades in the technology and defense industries, I have worked in both military companies and Silicon Valley. In addition, I have spent years tracking technology in China and getting to know the players in Chinese industry. I have traveled to China almost fifty times in the last fifteen years as a consultant to American and foreign companies, and have given countless presentations on Chinese high technology. I have also published three books related to China, including *China's Globalization (How China Becomes Number 1)*, which was a best seller there.

It is clear to me that China's leaders understand the benefits and threats of AI in a global sense. They know that leadership in AI is far more important than leadership in nuclear weapons, because, like steam and electrical power, AI will impact all industries civilian and military. What is more, the Chinese government's support for AI is extraordinary in its breadth and character. It provides funds to develop AI technology on the cutting edges of human endeavor, such as quantum computing and quantum communications. The Chinese government provides funding for developing next-generation robots and for the development of the markets that will buy them. The result is the simultaneous creation of both the sellers and the buyers of new innovative machines. In addition, the Chinese are developing AI-empowered trains that will run at as fast as 1,000 km per hour, as well as AI to improve crop yields and AI in small modular nuclear reactors designed to reduce pollution. Already China leads the world in broadband communications, digital currency, and three-dimensional facial recognition—technologies that all use AI applications.

AI-empowered technologies also have a dark side, at least from an American perspective. By 2040, smart "superphones" will transmit information to the cloud, reporting the moods and physical conditions of people, who can be monitored in real time by AI algorithms. Among other things, this helps the government search for dissatisfied citizens who are likely to become dissidents. An efficient cashless payment in the near future will also feed data into China's AI-enabled "social credit system" that monitors and tracks all purchases throughout the economy. China today has already deployed its social credit system nationwide in order to rank individuals and businesses based on their behavior and purchasing patterns. These social credit scores are then used to impose punishments on people with low social scores—such as restricting their ability to travel, preventing them from holding certain jobs, and even making it impossible for them buy certain goods.

The effective implementation of 3D facial recognition can dramatically increase a nation's economic competitiveness and help improve national security in ways that are compatible with individual liberties. But facial recognition technologies can also give the state the power to monitor the physical and mental health of a population by analyzing facial expressions and other markers. In essence, the state can monitor the satisfaction levels of great swaths of their citizens.

Based on the research that has been done for this book, I can say there is a high probability that by 2040, China will be in a dominant position in many areas of AI. It is, consequently, important for the leaders in the United States and other countries to understand the benefits and threats of AI and what actions are necessary to achieve AI leadership.

What has been the reaction of the United States government, industry, and public to AI development in China? Considering the scope of the challenge, the answer is very little.

Certainly, American companies have led in AI innovation. No corporation today has done more innovative research than Alphabet's Google. That company is indisputably the world's leader in the AI field. Facebook (now Meta), Microsoft, Amazon, Nvidia and many other smaller companies are also doing remarkable work. But in terms of the systematic implementation of AI to impact society, what China is doing is quite different. From China's start-up sector to its high tech giants, from academia to its military, one can see the hand of government spurring the development and execution of AI initiatives designed to revolutionize the economy and the nation's competitiveness. In the United States, by contrast, government leadership and support in recent decades has been next to non-existent.

It has not always been this way. The United States once was superb when it decided to take on mighty technical challenges. For precedents, one can look at Franklin D. Roosevelt's mobilization for World War II and Dwight D. Eisenhower's initiatives to buttress the national defense in the 1950's with the assistance of the best administrators and scientific minds available. As part of that effort, Eisenhower spurred the creation of the interstate highway system. One can also look at John F. Kennedy's commitment to the space race. These initiatives not only rapidly achieved their goals, they spawned a host of technologies that made possible, among other things, the microprocessor, the internet, smart phones, and GPS.

So what should the United States do now to counter the AI threat of China? The answer is straightforward: make investments to ensure leadership in technology while letting entrepreneurs innovate. Funding is required to build structures that help companies operate efficiently within their respective areas of expertise, whether in automobiles, medical devices, or computer chips. Attempts by the United States government to slow down China are short-sighted and simply will not work. Rather, support for innovation and implementation of AI at home will bring maximum dividends for America and the world.

This book provides granular details on industries impacted by AI and provides a look at the challenges and opportunities now facing the United States. The AI race will require sprinting in some areas and running a marathon in others, but it starts with the recognition that what we do in the next ten to twenty years will be critically important to the future economic and political health of the nation. The optimum situation is for the United States and China to cooperate on AI. Unfortunately, this would be exceedingly difficult to achieve within the present political environment. Therefore, direct competition is the best option. To compete, it is important for leaders in the United States to understand where China and the United States are in the race to create a smarter planet. Hopefully, this book will help begin that process.

Chapter 1

Miracles and Dangers: The Coming of AI

The AI Revolution

In the year 2040, you will never be truly alone. Never will you grope for a lost word, a forgotten name, a fact hidden deep in your memory. There will always be a voice to communicate with you, in a virtual whisper, most of the things that you've ever known and all those things you will never forget. This voice will empower you, just as the combustion engine gave man the power to move faster than a cheetah. In the hands of a malicious government, however, it could also control your thoughts and your life as though you were a puppet on a string.

The name for this technology, powered by artificial intelligence (AI), is the virtual digital twin. Located in the cloud or an extremely powerful superphone, the virtual twin will contain an enormous store of data about you, including all the data gathered by sensors in your environment and information you have learned throughout your life. It will be instantly accessible to you, so that you will sometimes forget exactly what information came from your brain and what information came from your virtual twin. To other people, you and your virtual twin will seem nearly indistinguishable. At your job, for example, your employer will not just rate the work you do alone—you'll be judged on the combined output of both you and your virtual twin.

A person's digital twin will be a digital extension of both the person's data and his or her analytic capabilities. When a person writes a letter, makes an audio or video recording, or takes a picture, the digital twin can glean important data from that creative product and save it. Over time, the digital twin will also take data from social media or gather it with the help of the latest virtual reality capabilities. And as the twin gains access to more data, its capabilities will improve, and its powers of intelligence and analysis will grow ever stronger.

From the time a person acquires a digital twin, its memory will continually expand until an exhaustive database on a person has been established. That will include data about their families, friends, and communities. The twin's machine memory will mirror human memory, and data will be stored based on a specific task rather than in a mass of general information. The twin will also be able to initiate a search for data in the cloud. Unlike human memory, the data bank of a digital twin will never deteriorate. When people die, their digital twins will live on, leaving open the question of who will control the dead person's data. Together, the computing power and data bank of the digital twin, along with the big data centers it communicates with, will be able to perform most of the tasks that are normally done by the human brain. As twins come to populate the world, they will transform both the work lives and the private lives of every person who can afford one.

Like many machines powered by AI, the most advanced and powerful digital twins will come from one of two places: either the United States or China.

The digital twin will be one of countless manifestations of the AI revolution, a revolution that, for better and worse, will touch every nation, every industry, and every aspect of daily life. It is tempting to compare the AI revolution to two past technological revolutions that changed the course of human history: steam power and electrical power. But the AI revolution will be different in at least one major way—it will happen much, much faster.

Many emerging technologies promise great change and great benefits. Battery tech, for example, has the potential to reduce carbon emissions and fight global warming, and gene tech like CRISPR may eliminate many hereditary diseases. AI might appear to be just another powerful technology among a panoply of important advances. But to see AI in this limited context is a serious mistake.

AI is a master technology—a technology that enables and empowers nearly every other technology currently in development around the world. AI and machines that can learn on their own are the key to the future of innovation. As a result, a nation that leads in AI now is likely to become even more of a leader in a range of new technologies in the years to come.

In the near future, AI will transform warfare. Naval fleets will increasingly consist of unmanned ships and unmanned submarines, while rockets will be guided by AI-empowered computers. Clouds of tiny drones will move through the sky like flocks of birds, overwhelming their targets through the force of sheer numbers.

Meanwhile, the world's highways will no longer be clogged with traffic because AI will effectively direct it and far fewer vehicles will be needed. With the help of AI, a reduced number of electric, self-driving vehicles will be operated by ride-sharing services, quietly delivering passengers and goods from place to place without polluting the air. Running twenty-four hours a day without the need to park, they will free up vast tracks of land previously used for parking spaces.

AI-empowered virtual reality (VR) and augmented reality (AR) will let an architect enter and inspect a building that has not yet been built. And in healthcare, AI will dramatically speed up the process of discovering new drugs and individualized therapies, while drastically reducing their costs. People will live longer and should be healthier because AI will help determine their diets.

AI in China

China threatens to surpass the U.S. in the near future through strategic government-led and funded initiatives designed to make China the world's leader in AI and other key, cutting-edge technologies. If that happens, the United States will be destined to become a second-rate power. Without a major public policy shift, the AI gap with China will grow to insurmountable proportions, with a disastrous impact upon the United States' defenses, its economy, and the quality of life of its citizens.

In 2017, the Chinese State Council unveiled a plan to become the world leader in AI by 2030. It declared that its aim was to surpass all rivals technologically and build domestic industries in everything from agriculture to medicine to manufacturing based on big data and AI. The top rungs of the government said a massive effort would take place to ensure its companies, government, and military leap to the front of AI development. The multibillion-dollar initiative was, according to the State Council, a key impetus for economic transformation. Later that year, the Chinese central government declared self-driving vehicles, smart cities, and medical R&D (research and development) as key platforms for development.

What is now clear is that China has made enormous progress in AI since that announcement. It is poised to take the lead in artificial intelligence applications for manufacturing, health care, and transportation, leapfrogging the West in key new technologies.

For example, China is well on its way to building ten million 5G mobile base stations by 2025, essentially wiring the whole country for game-changing AI technologies. With 3D facial recognition, people in China can board trains, buy produce, and obtain medical services without physical contact. In the next few years, China is expected to deploy an advanced medical system that uses AI, big data, and sensors. Key to this new health system will be robots, which will rely on big data to conduct various tasks, including elderly home care. Autonomous transportation will require the coordination of AI-empowered automobiles, buses, local trains, high-speed trains, and other forms of transit. Entirely new cities are being built to support such autonomous transportation. And a new generation of drones is under development for goods transport.

"China stands *today* as a *full-spectrum peer competitor* of the United States in commercial and national security applications of AI," Eric Schmidt, the former CEO of Google, now called Alphabet, wrote in 2020.[1] "China's advantages in size, data collection and national determination have allowed it over the past decade to close the gap with American leaders of this industry. It is currently on trajectory to overtake the United States in the decade ahead."

Behind these developments stand at least three structural advantages that favor China.

First, China has an AI industrial policy that puts government in partnership with the private sector to spur development. This cooperation helps accelerate the adoption of new AI technologies in the real world at a much faster pace than would be possible in the United States. In the U.S., the federal government and Silicon Valley function quite separately, often divided by economics, philosophy, and culture. In China, the central government has the ability to muster unified national, regional, and local resources on behalf of AI development. For example, as the Chinese tech giant Baidu builds autonomous cars, the company is working with municipal governments to remake cities to accommodate the self-driving vehicles. Those cars are already operating in a handful of Chinese cities. In the near future in China, a Baidu executive recently said, Chinese cars will be guided by laser-based lidar sensors, radar, and cameras. There will also be street signs designed just for these sensors. It is not hard to imagine, therefore, China becoming a nation of self-driving cars far sooner than America.

A second advantage for China is data. What labor and equipment were to economies in the industrial era, data is in the digital era. The open-source nature of AI research in the United States makes its technological advantage almost irrelevant in the long run. It is easily available to Chinese scientists. What will be more important is having access to vast troves of data. And China has more high-quality data than anyone else.[2] There are close to one billion people online in China, three times the number of Americans who are online. When those millions search, watch videos, and make payments on their mobile devices, they leave an immense, rich data trail.

Controlling COVID-19 has also generated a large amount of data, and China is using this expertise to monitor other health activities.

Third, a new wave of scientific talent is aiding China's AI development, with China's universities graduating three times as many computer science graduates than the United States[3]. In addition, a significant portion of computer science and STEM (science, technology, engineering, and mathematics) graduates in the United States are Chinese nationals, who will take their American training back to their homeland.

China's AI development is more than a matter of government edict. The AI private sector is a teeming, highly competitive, and often unregulated ecosystem that includes thousands of struggling start-ups and mid-sized growth companies as well as a handful of reigning behemoths. Just as the giant American corporations Alphabet, Facebook, and Microsoft have led in research and application of AI in the United States, the largest Chinese companies—such as Baidu, Alibaba, and Tencent—have made huge investments in AI and are now seeing marked results. In 2017, the central government split the first wave of open AI technologies between Baidu, Alibaba, and Tencent. It assigned Baidu to the development of self-driving cars, Alibaba to smart cities, and Tencent to digital healthcare. The government has recently focused on many more companies and many new high-growth companies are emerging.

China 2040

One metaphor for the future trajectory of AI is the progress of the aeronautical industry over the last century. The Wright brother's first successful airplane flight, in December of 1903, lasted a total of twelve seconds and covered a distance of only 120 feet, which is considerably shorter than the 200 plus foot wingspan of the modern Boeing 747[4]. The Wright brothers' first airplane, the Wright Flyer, was controlled by primitive levers and pulleys[5]. By contrast, the Boeing 737 contains over forty miles of wires to carry information from flight avionics and other electronic systems[6].

As any modern business traveler knows, the technological complexity of modern aircraft is an astonishing feat of engineering. Over decades of incremental engineering improvement, the modern aircraft has developed into a platform that has made the world smaller. We now find ourselves at a similar inflection point, analogous to the moment experienced by the Wright brothers a little over a century ago in Kitty Hawk, North Carolina, as AI begins to take flight.

While it is still relatively early on in the AI adoption curve, in 2021, China has already broadly adopted AI technology in industries such as defense, healthcare, transportation, and telecommunications as well as in applications such as virtual reality, image recognition, logistics management, and digital currency. In many cases, the Chinese government consciously creates the markets that will use new technologies years in advance. And more cutting-edge technologies are coming.

Chinese high-tech companies are accelerating technological developments at an extraordinary pace, in part because the country is still in a relatively early stage of its industrialization process and partly because of the funds flowing from government sources. Many applications that use AI in China do not have to compete with legacy technologies. For example, electronic payment by smartphone was easier to implement widely in China because the nation never really adopted credit cards as a means of payment. By contrast, in the United States and Western Europe, older technologies are well entrenched. These legacy technologies in the U.S. and elsewhere slow down the adoption of AI because of the power of bureaucracy and the continued financial gains from existing technologies.

By the year 2040, China's urban majority will find itself immersed in AI and big data. While residents of China's rural provinces will have significant contact with AI through its use in agriculture, AI technology will penetrate nearly every aspect of urban citizens' lives.

Imagine for a moment we are in the Nanshan tech district of Shenzhen in 2040. It is very early on a weekday morning, and we find ourselves standing on a monorail platform. It's so early that the mass of the city's workers has yet to gather on the platform for their daily commute into work, and the last of the street cleaning robots and overnight service drones are still finishing their autonomous tasks from the night before.

As the sun rises, self-driving cars become visible below. Their electric motors hum from inside windowless chassis as they ferry to work those who can afford to hire them for the journey. Other less affluent people will ride AI-controlled scooters. Beneath the platform, the makeup of the storefronts has changed as well; with a decrease in demand for in-store shopping and sit-down restaurants. Shenzhen, like other cities in China, has converted shopping centers, department stores, and restaurants to new uses, such as leisure centers or medical diagnostic clinics. Housing dentistry can also be increased.

The streets of Shenzhen, like those of other major Chinese cities, are arranged differently than they are today. In 2040, the road network is divided between pedestrian walkways and separate, dedicated roadways for self-driving vehicles and robotic transport, which helps to smooth the flow of traffic between humans and autonomous vehicles. Camera surveillance systems, connected to AI-based, three-dimensional facial recognition technology, provide security and monitoring in this citywide digital panopticon. By using 3D facial recognition, the transportation system eliminates the need for train tickets and entry is seamless.

As workers arrive and crowd onto the high-speed trains, most are glued to their 6G "superphones," which allow them far greater connection speeds and throughput than was possible with the fastest internet landlines just a generation earlier. The phones not only stream high-speed data and video, but, along with wearable wireless devices, also are a key access point for AI content and services. Other workers wear VR headsets, passing their time in their alternate realities.

This two-way data connectivity, provided via a 6G wireless backbone, allows AI capabilities in enormous data centers to be accessed remotely across a wide geographic area. In 2021, China had already built a network of these strategically located data centers in domestic locations as well as abroad; some of the data centers are buried deep underground, while others above the surface are so large that they are visible from space. A benefit of such technology is superior brain power and the ability to manage civilian industries and military ecosystems.

In China's major urban areas in 2021, cell phones were ubiquitous. Chinese city planners and civil engineers were already developing their next-generation transport infrastructure. But the changes that AI are driving in 2040 are far more profound.

In 2040, AI-based medicine is dramatically increasing average human lifespans. Longer lifespans can help improve economic productivity by adding additional decades to the average worker's career in the workforce.

Higher-density cities with clean air are possible as a result of the convergence of many technologies driven by AI. Due to the centralization of businesses and services, as well as better job opportunities, far more people live in large cities. In 1950, New York City and Tokyo were the first megacities, with populations of more than ten million people. By 2021, the world had thirty-three megacities with even larger urban populations; greater Tokyo, as an example, in 2021, had a population of over thirty-seven million people. By 2040, there will be more than fifty megacities, some of which are expected to accommodate as many as fifty million people.

When residents of Shenzhen purchase their goods and services in 2040, they are able to pay for them using 3D facial recognition technology, which identifies them and then programmatically deducts the purchase price from their digital bank accounts. These new digital payment architectures are capable of supporting hundreds of millions of digital transactions in real time.

China became the first country to adopt digital currency on a nationwide level in 2021, which has vastly simplified commerce and reduced the ability of citizens to hide cash. When the digital yuan was first rolled out in 2021, many transactions were still conducted in cash. By 2030, however, Chinese banknotes and coins are likely to be obsolete and no longer a viable payment option for any of China's citizens.

Digital currency and financial transactions are both closely tied to China's 3D facial recognition initiatives. In 2021, two of China's largest fintech players, Ant Financial Services and Tencent Holdings, used facial recognition technology at point-of-sale terminals in limited pilot projects.[7] By 2040, the technology is almost everywhere.

(The Chinese government in 2021 was, however, concerned with private institutions having control over financial transactions and creating their own digital currency, and so the actions of Alipay and WeChat Pay were being restricted. Also, China banned Bitcoin and other digital currencies because of concerns with illegal transactions.)

Also in 2040, citizens in China's cities have a wide variety of sensors connected to them, generating a great stream of data. In 2021, automobiles, robots, smartphones, and other digital devices generated large amounts of data that was growing exponentially. By 2040, humans generate a dense trail of data wherever they go, analyzed by AI tools. A large amount of that data will be stored in the enterprise cloud, where it will also be analyzed by AI. The cloud will have many layers, including a personal cloud that will follow individuals in real time, and each layer will have its own AI capabilities.

The intelligence generated by this new information stream will also be complemented by super-bodies enhanced with mechanical-digital exoskeletons. Like their cognitive counterparts, these exoskeletons, which will also utilize AI and sensor data, can significantly expand humans' physical capacities. Even in 2021, there were credible reports of the Chinese military using exoskeletons in the field to conduct patrols in the harsh, high altitude conditions of their Himalayan border regions.[8]

The Dark Sides of AI

Then, there are the potential dark sides to AI. We can see that too when we look more closely at Shenzhen in 2040.

The superphones that people are holding on the train platform in 2040 during their morning commute transmit information back to the cloud, reporting the moods and the physical conditions of the people, who can be monitored in real time by AI algorithms. Among other things, this helps the government search for dissatisfied citizens to determine who are likely to become dissidents.

The efficient cashless payment system also feeds data into China's AI-enabled "social credit system" that monitors and tracks all purchases throughout the economy. Two decades earlier, China deployed its social credit system nationwide to rank individuals and businesses based on their behavior and purchasing patterns. These social credit scores are now used to impose rewards and punishments on people with low social scores—such as restricting their ability to travel, preventing them from holding certain jobs, and even making it impossible for them buy certain goods.

The effective implementation of 3D facial recognition can dramatically increase a nation's economic competitiveness and help improve national security in ways that are compatible with individual liberties. But facial recognition technologies can also give the state the power to monitor the physical and mental health of a population by analyzing facial expressions and other markers. In essence, the state can monitor the satisfaction levels of great swaths of their citizens. By knowing every individual's feelings with precision, the state gains the ability to use coercion to change the behaviors of those it deems a threat to the social order.

In 2040, Shenzhen's increased worker productivity clearly creates economic benefits for the city and its citizens. But the government faces the challenge of keeping its citizens productive and employed throughout longer working lives. Large-scale displacement of jobs by smart robots and AI require the Chinese government to provide a large number of unemployment benefits to an urban workforce experiencing structural unemployment. For example, most predictable, labor-intensive, industrial tasks are performed in 2040 by autonomous robots. Service robots can be seen scurrying across cities, delivering goods and food, running errands, fetching laundry, picking up trash, and cleaning streets, walkways, and parks.

What will large numbers of unemployed workers do to stay engaged while they are living on government assistance?

In 2040, many of the unemployed Chinese retreat into virtual reality worlds to occupy their time. VR addicts no longer travel extensively to different geographic regions—or even locally to museums, bars, and restaurants. Instead, they only travel by way of virtual reality that provides immersive and nearly lifelike experiences.

Large companies and the government use VR-based promotions and propaganda to further their views and direct public policy. Among the younger members of society, AI and VR-related activities, including education and possibly gaming, are essential parts of their daily routine.

Some individuals even withdraw completely because they no longer feel mentally, intellectually, or emotionally connected to the real world. VR addicts only need a small living space with minimal physical possessions and may acquire virtual possessions online. For the unemployed and those who no longer wish to work, the virtual world is a kind of digital sanctuary, well insulated from the pressing concerns of reality.

(In 2021, China restricted access to gaming for minors, limiting gameplay for children under eighteen years old to one hour per day on weekends and vacations, and eliminating gaming entirely on school days. However, this regulatory regime may run counter to a broader trend that will see the Chinese embracing online virtual worlds in the years to come.)

In 2040, as many citizens withdraw from society, AI creates a bifurcation, a digitally divided society, split between highly trained workers, who program and manipulate AI and its adjacent technologies, and those non-technical workers who do not. Into the latter category will consist of part of the population that simply uses AI technologies as consumers and doesn't possess a technical understanding of its inner workings. But most new businesses that utilize AI and its associated technologies will need workers with a deep knowledge of AI computer programming, such as highly skilled engineers. These workers are needed to modify AI for superphone applications and other platforms. People without those skills will be left behind.

The Clash of Two Systems

Exploring the development of AI in China and the United States provides a prism through which one can observe two very different economic and political systems at work: One system is run according to the teachings of Xi Jinping, another system follows the tenets of Milton Friedman. Which system succeeds will be heavily impacted by the role of AI and will tell us a great deal about the political, economic, and military future of the human race.

AI is power. It will empower nations to dominate commercial markets and influence economic and political life in all regions of the globe. And the nation that leads in AI will control the most valuable asset of this new era: data—which will be the lifeblood of an increasingly smart planet.

The Chinese government's long-term development goals bring together universities, think tanks, the nation's private tech giants, government-controlled corporations, the military, and various other branches of government to reach goals that are related to AI. Beyond the sheer mass of funding and organization involved, what is impressive to an American observer is the Chinese willingness to absorb losses and setbacks in the short- and medium-term to achieve critical long-term goals. For example, the Chinese Communist Party and the government decided to invest $100 billion to build fifth generation, or 5G, wireless networks for the transmission of data when markets for that technology were still in formation and profits from those investments were still far off. There will be an even larger investment in 6G, leaving the United States and its allies even further behind.

By contrast, the U.S. has preferred to let the private sector lead the charge in the development of critical technologies like AI and has placed restrictions on its use. The result has been America planning based on shorter-term market forces and shorter timelines than China. It has also meant a failure to make the kind of government investments in industries and infrastructure that the Chinese have made. The reduced role of U.S. government leadership in the development of key technologies since 1980 stands in sharp contrast with policy in the mid-twentieth century.

Government-led initiatives after World War II gave birth to the core-enabling technologies of our twentieth and twenty-first century economies—from the interstate highway system to the transistor and microprocessor to the internet.

In the past decade, the Chinese Communist Party has demonstrated that China has the technical prowess, the political will, and the organizational ability to mount one major national, high-tech initiative after another. During that time, the American government's response to technological challenges have been hesitant and sclerotic, particularly in the realm of defense.

Look at the divergent responses of China and the United States to the rise of COVID-19. The Chinese were far ahead of the U.S. on their use of data-driven contact tracing, disease testing, and mass vaccination. While the West would not and should not engage in all the techniques used by China, the Chinese were, in the end, far more successful at containing COVID-19 than the United States.

The ancillary impact of the COVID-19 crisis also speaks volumes about the United States' ability to plan and execute complex public policy responses. In the fall of 2021, for example, there were seventy cargo ships off the California coast waiting to be unloaded and to have their cargo distributed across the United States. Due to COVID-19-related labor market constraints, there were not enough workers in places like Long Beach and San Pedro, California to unload the goods from cargo ships. Without truck drivers and forklift operators to facilitate the transit of goods to warehouses and then on to the distribution centers of Walmart and Amazon, the United States was suffering from major supply chain disruptions. Neither the private sector nor the federal government had a plan or an ability to contend with this crisis. China, on the other hand, was never plagued by a similar supply chain challenge in the wake of COVID-19. It responded rapidly to logistics problems in Ningbo and disruptions were kept to a minimum.

The United States, during the Biden administration, has made some attempts to increase the investment in its critical domestic technology infrastructure. For example, in June 2021, the U.S. Senate passed the U.S. Innovation and Competition Act, which included $52 billion in federal investments for domestic semiconductor research, design, and manufacturing. But in the months that followed, Washington dithered.

Semiconductor manufacturing capacity in the U.S. had eroded from 37 percent of global capacity in 1990 to 12 percent in 2021. This was mostly because other countries' governments invested ambitiously in chip manufacturing incentives and the U.S. government did not. American companies failed to make adequate investments as they sought to focus on short-term financial gains.

While this legislation at least acknowledged the gravity of the United States' challenges in semiconductors, the challenges related to AI are much greater and not as transparent. This one action should not be confused with a comprehensive and coherent overall strategy for the nation. The United States simply has no master plan—no effective policies for developing domestic high-tech capacity in general and, most critically, in the application of AI. And the private sector, governed by market forces alone, cannot meet the challenge, especially when the United States government is trying to limit some activities of companies particularly innovative in AI.

To be sure, the American tech industry has had huge successes in the twenty-first century, especially in terms of innovation, creativity, content development, and marketing. A nation that developed the iPhone, Google search, Amazon, and Tesla is certainly doing something right.

The key issue is how will the United States lead in the application of AI and other critical technologies in the longer-term—a competition between American industry on the one hand and the Chinese government together with private industry on the other.

What is at Stake

As China continues its inexorable march toward becoming a global superpower, American leaders must understand the magnitude of the threat that the United States faces. China is a significant threat across multiple domains. It's not hyperbole to say that a rising, AI-empowered China represents an existential threat to the United States' global leadership along military, political, and economic lines.

Falling behind China significantly in AI could impact and leave vulnerable all areas of the U.S. military on land, sea, air, and space. Wherever data plays a role in defense, America has a vulnerability. With a vast increase in the flow of data in battle zones, commanders on the bridges of ships and land-based command centers will become choke points of information, unable to make and execute decisions as quickly as their AI-enabled adversaries. The result could be devastating to American forces. Aircraft carriers and carrier battle groups could be defeated and/or left useless by the increasingly advanced development of Chinese drones and unmanned aircraft, ships and submarines, and other AI-empowered technologies. This scenario would be very likely to substantially degrade or eliminate America's ability to project power around the world.

The U.S. Navy is behind China in its ability to launch unmanned vehicles off carriers, limiting their range. Aircraft carriers are particularly vulnerable to drones and may risk destruction by unmanned vehicles in the near future. Inferior AI capabilities make communications vulnerable, potentially leaving U.S. ships unable to communicate with each other or with their command centers onshore. American radar and other key technologies could be blocked, leaving a fleet blind and unable to function.

At American Pacific bases, U.S. Marines and other land forces could be extremely vulnerable to air attacks. Without carriers and carrier battle groups, the United States loses the ability to keep open crucial shipping lanes of the South China Sea, and therefore the capacity to come to the aid of its allies in the region.

Since the mid-1940s, the world's economic and political structure has been reliant on the military and financial support of a powerful United States. The loss of American military power and economic hegemony would jeopardize relationships between nations and organizations—altering or ending an international system that has largely kept the peace and allowed unprecedented global economic growth for nearly eight decades.

The United States has maintained a special place at the heart of the global monetary system since the Bretton Woods Agreement of 1944. The agreement pegged the value of the dollar to gold and established a regime of fixed exchange rates against the dollar for the currencies of the world's major industrialized nations—it—placed the U.S. dollar at the epicenter of the international monetary system. Even after President Richard Nixon removed the United States dollar from the gold standard in 1971, the dollar today remains the global reserve currency, with 59 percent of global central bank currency reserves denominated in dollars.[9] The benefits to the United States are manifold. One obvious example is in securing natural resources—since virtually all oil traded around the globe is still quoted in U.S. dollars. Moreover, the U.S. dollar's reserve currency status has direct national security benefits. The United States controls the underlying plumbing and mechanics of global dollar payment systems, such as the SWIFT messaging system, which are profoundly helpful to the U.S. in furthering its economic objectives against nation-states the U.S. government designates for economic sanctions. If the U.S. dollar were to lose its reserve currency status, the capacity of the United States and its government to project economic power around the world would be severely diminished.

With the United States weakened, Japan, South Korea, Singapore, and Australia would become far more vulnerable to economic and political intimidation by China and may become part of a Chinese sphere of influence. Moreover, China currently claims Taiwan as belonging to the People's Republic of China under its One China policy; a reduction in the effectiveness of U.S. forces in the region would increase the chance of China taking possession of Taiwan by force.

In a worst-case scenario, as with the diminution that befell Great Britain in the wake of the Second World War, the United States could become a "smaller," less significant nation with a lower standard of living. Reduced income and living standards would likely contribute to increased political polarization, weakened American political institutions, and greater political instability.

The challenge from China is far greater than the Cold War rivalry with the Soviet Union ever was. For the first time since the United States became an industrial nation, it faces a competitor with the potential to have a larger economy, a larger population, a peer-level military, and an ability to develop cutting-edge technologies with a speed and competence as great or greater than the United States.

There is nothing preordained about China surpassing the United States in AI and, as a result, surpassing the United States as a great power. But to assure that does not happen, America must act now.

The Evolution of Machine vs. Human Power

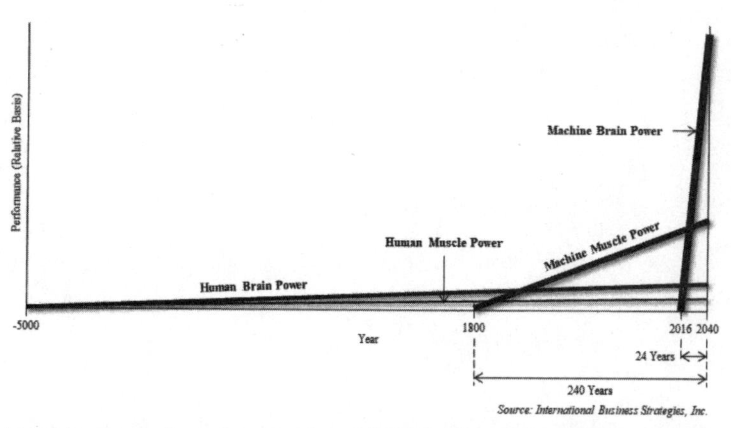

Machine muscle power, such as that produced by steam, electric, and gasoline engines has had a dramatic effect on society over the past two hundred years. But AI, or machine brain power, will have a more profound impact between 2016 and 2040.

Chapter 2

Who Is Leading the AI Revolution?

In 2017, Eric Schmidt pitched Google's new AI product to Chinese technology executives. As the brilliant leader of the world's leading AI company, he may have expected a warm reception. If so, he was very wrong.

At a meeting in Wuzhen, China, Schmidt, the chairman of Alphabet, Google's parent company, argued that Chinese tech giants like Alibaba, Tencent, and Baidu needed TensorFlow—the new Google AI software that could recognize objects, identify spoken words, and translate between languages. Schmidt said that TensorFlow represented a revolution that could reinvent China's tech sector. Google's software would allow Chinese companies to perform tasks that could increase corporate profitability, such as highly targeted online advertising and automated decision-making for consumer credit. Schmidt argued that TensorFlow would be a win-win for everyone involved, expanding markets for the American tech giant while simultaneously adding cutting edge artificial intelligence capabilities to China's corporate sector.[10]

What Schmidt did not say was how Google would benefit. TensorFlow was the successor to the massive software system that trained deep neural networks across Google's global chain of data centers. The software was open source, so if other companies, universities, or governments were to use Google's software as they pushed into artificial intelligence, those efforts would feed and enhance the capabilities of Google's research efforts. Most of all, if TensorFlow became the de facto standard for building artificial intelligence, Google believed it could bring the world to its cloud computing services—a critical area of growth for the company.

But as Schmidt spoke, many in the audience that day in Wuzhen sat stone-faced. The truth was the Chinese would never consider adopting TensorFlow. China's tech giants were already racing towards an AI-enabled future, embracing the latest concepts in AI and deep learning, and planning to move forward independent of the United States. The Chinese had been building their own AI capabilities for years; like Google, China's tech giants were erecting a vast network of specialized machines to collect data and develop applications.

Even if China's tech sector had not been busy playing catch-up across a broad range of technologies for years, the Chinese leadership was committed to the idea that if there were ever to be global AI standards, they would not be standards created by American companies. The standards would be Chinese.

"I knew when I gave the speech that the Chinese were coming," Schmidt later said. "I did not understand at the time how totally effective some of their programs would be. I honestly just didn't understand."[11]

While Google was ahead of Chinese companies in AI, the Chinese leadership was convinced that AI was a key part of the growth and digitization of China. There was no way China would depend on an American company for this crucial technology.

An American Story

But Schmidt could be forgiven for underestimating Chinese AI technology in 2017. Most Americans underestimated China's high-tech sector—and continue to do so to this day. In part, that is because most of the foundational research on AI during the twentieth century took place in the United States.

The history of AI stretches back more than six decades, with the earliest theorists performing their first pioneering experiments in the 1950s. The cornerstones of today's AI technology are "neural networks," a technology similar to neurons in the brain that permit a machine to learn on its own. Neural networks were first proposed in 1944 by Warren McCulloch and Walter Pitts, two University of Chicago researchers who moved to MIT in 1952. Modeled on the human brain, a neural network in a computer learns to perform a task by analyzing training examples. A computer might be fed thousands of images labeled as cars, chairs, tables, houses, and coffee cups, and it would find visual patterns in the images that consistently correlate with particular labels. Today, a neural network consists of thousands or even millions of simple processing nodes that are densely interconnected and layered.[12]

In the 1950s, however, computing power was not nearly great enough to make neural networks perform anything but the most primitive functions. In fact, the potential of neural networks was attacked by other scholars in the nascent field who wanted to pursue another approach to AI and who competed with neural network supporters for funding. Although there had been a wave of publicity in 1958 about an early machine called Perceptron that used neural networks, it was soon forgotten, and the technology fell into obscurity for decades.

A handful of scientists, however, kept the idea alive.

Geoffrey Hinton and Google

One of those scientists was Geoffrey Hinton, a member of the prominent British Hinton family, renowned for its brilliant and innovative thinkers. Hinton's great-great-grandfather was George Boole, the mathematician and philosopher whose "Boolean logic" provided the mathematical basis for modern computers; his great-grandfather was Charles Howard Hinton, a mathematician and fantasy writer, whose idea of a fourth-dimension informed science fiction for more than a century. Hinton's great uncle invented the jungle gym; a cousin was one of the few female nuclear scientists working at the Manhattan Project; and another cousin, Sir George Everest, the surveyor-general of India, had the world's tallest mountain named in his honor.[13]

As an undergraduate at Cambridge University, Geoffrey Hinton was interested in learning how the brain worked. He tried his hand at physiology, chemistry, physics, psychology, and philosophy, but those fields could not provide the answers he was searching for. The truth was science's conception of the brain at that time was simply too limited. In the end, Hinton left Cambridge to become a carpenter in London who spent endless hours following his interests in the local library. Looking back at earlier neural network work, Hinton was convinced that biological and artificial intelligence could help one another move forward. Eventually, he wound up taking a Ph.D. in artificial intelligence at the University of Edinburgh and, after post-doctoral work, secured a position at Carnegie Mellon University in the United States. Throughout this time, Hinton never lost faith in the power of the neural approach to AI and developed relations with fellow believers in Europe and the United States.

At Carnegie Mellon, Hinton showed that a type of artificial neural network could do something extraordinary: When he fed the machine part of a family tree, it could discern the relationships of other family members that it had not been given. Having accomplished that task, clearly it could do much more, both with images and with words. In short, the machine could teach itself.

The first practical application of Hinton's research came in 1987, when researchers at the Carnegie Mellon artificial intelligence lab were trying to build a truck that could drive itself. They were making very slow progress when a graduate student named Dean Pomerleau replaced its software with software inspired by a recent paper written by Hinton and a colleague. The truck could now learn to drive by watching humans navigate the road. At first, it could only drive nine miles an hour in a local park but, over time, its capabilities improved. On a Sunday morning in 1991, it drove itself from Pittsburgh to Erie, Pennsylvania at sixty miles per hour.

With interest in neural technology revived, Hinton moved from Carnegie Mellon to the University of Toronto, where he has continued his work on neural networks. But it was not until 2012 that Hinton had an earthshaking breakthrough. He and two graduate students built a neural network modeled on the brain that changed the way machines saw the world. The machine could recognize real-world objects with an accuracy and power never before achieved. Moreover, it could learn human-like skills by analyzing vast amounts of data. Hinton called this "deep learning." Its potential was majestic. It could someday transform not just computer vision but everything from talking digital assistants to driverless cars to drug discovery.

As the use cases for deep learning proliferated, the world came knocking on Hinton's door. First came the Chinese technology company Baidu, which offered Hinton and his two students $12 million to work for the company for several years. But the three scholars instead chose to form a new company, DNNresearch, and sell it to the highest bidder in an auction. Baidu, Microsoft, Google, and a London-based start-up named DeepMind began the bidding.

In the end, Google purchased Hinton's company for $44 million, allowing Hinton to continue his work part-time at the University of Toronto.

Demis Hassabis

DeepMind, the start-up that competed with Baidu, Microsoft, and Google in the bidding for Hinton's company, was nothing if not ambitious. Its founders intended to create "artificial general intelligence" or AGI—machines that could do anything the human brain could do, only better. The company's lofty ambitions were driven in large measure by one of its founders, Demis Hassabis, the son of a Chinese-Singaporean mother and a Greek Cypriot father who ran a toy store in London. Hassabis had three passions: computer games, artificial intelligence, and winning at both of them.

Hassabis had been rated the second-highest, under-fourteen chess player in the world. At twenty-one, he entered the Pentamind competition in which players from around the world competed at five games of their choice. Hassabis won that year and in five of the next six years. (The one year he did not win was a year in which he did not compete.) After graduating with a First (First-Class Honors) in computer science from Cambridge, Hassabis helped design one of the world's most popular—and revolutionary—computer games, Theme Park, in which players build and operate a sprawling amusement park. The game spawned countless imitators and the new field of "sims" games. Later, with a fellow graduate from Cambridge University, Hassabis founded a game company called Elixir Studios and then pursued a neuroscience Ph.D. at the University of London to deepen his understanding of the brain.

During his post-doctoral work at the university, he met a fellow researcher, Shane Legg, who shared what was then considered to be the outlandish goal of "solving intelligence" and creating AGI. Legg wanted to pursue this goal in academia. Hassabis argued that they could never attract the necessary capital from grants in academia to build the equipment that was needed. With one start-up behind him, Hassabis argued that establishing a private company with investors was the only way to succeed. Together with a social activist and entrepreneur named Mustafa Suleyman, Hassabis and Legg founded DeepMind in 2012, naming it after the supercomputer designed to calculate the ultimate question of life in the science fiction novel, *The Hitchhiker's Guide to the Galaxy*. Its goal in the longer term was creating AGI, although it would also create nearer-term technologies. One of DeepMind's first investors was Peter Thiel, a founder of PayPal and an early investor in Facebook, LinkedIn, and Airbnb. Another was Elon Musk, the PayPal veteran who went on to establish Tesla and SpaceX.

Like Hassabis, DeepMind was immersed in computer games and artificial intelligence. In fact, DeepMind used games like Space Invaders to train AI systems. As they played the games,

AI systems figured out what strategies earned the most rewards and eventually became extraordinary players. This technique—called reinforced learning—could be applied outside of games. A robot could learn its way around a room, or a car could navigate a neighborhood. Reinforced learning could also be used to teach a machine the English language.

After Google had secured the services of Geoffrey Hinton in 2012, Google's cofounder, Larry Page, authorized the hiring of any and all promising researchers in the field. The next year, a team of Googlers, including Hinton, traveled to London to check out DeepMind and its work. They found a company with brilliant potential but with no revenue to speak of and little ability to pay competitive wages to its all-star team of researchers. Although Hassabis had promised that DeepMind would remain independent for twenty years, the reality was that the start-up did not have the deep pockets necessary to reach its objectives. Without financial help, the company would die.

Google purchased DeepMind for nearly $600 million, but the deal had a number of unusual aspects. DeepMind would remain in London and would not become part of Google's main AI entity, GoogleMind. Google was barred from using DeepMind technology for military purposes. And Google was required to create an independent ethics board that would oversee DeepMind's AGI technologies.

A New Alphabet

The acquisition by Google of Hinton's company and Deep Mind eventually launched Google and Alphabet, its parent company, on a new trajectory. AI and neural networks would come to define what Alphabet was about.

In a keynote presentation in 2015, Alphabet's CEO Sundar Pichai outlined the importance of artificial intelligence to the technology world and to the company in the years to come: "It's clear to me," Pichai said, "that we are evolving from a mobile-first to an AI-first world."[14]

Since then, AI has been a thread that runs through search and advertising, cloud computing, autonomous driving, healthcare, gaming, and bets the company has made in other fields. The company has been rethinking all of its products for an AI-enabled future, and it has focused on developing sophisticated machine learning capabilities through both outside investments and in-house development.[15]

Today Alphabet is America's undisputed leader in both research and development of AI technologies. Its AI research work partly resembles that of an academic institution, with hundreds of scientists studying concepts years away from immediate use. But the company is also massively involved in applying AI to current and future products, services, and acquisitions. The company invested $26 billion in research and development in 2020, most of which went to AI research.

Google's strategy includes protecting its core revenue sources—search and advertising—which is now under competitive pressure from both Amazon and Facebook. To do so, it is positioning itself to dominate adjacent sectors such as digital commerce, branded hardware products, and content. It is also integrating its services into every aspect of the digital user experience. In addition, Google is looking to use its lead in AI to disrupt healthcare, transportation, logistics, and other industries.

Central to Google's growth plan is expanding its cloud computing business and its related software, TensorFlow. This open source AI library uses data to build models for outside developers to create large-scale neural networks. Those networks have many layers that support data analysis. Millions of software developers are creating AI-based applications based on TensorFlow because of the wide range of specialty applications that can be addressed. TensorFlow supports various programming languages, including Python, C++, JavaScript, Java, and Go. Additionally, TensorFlow is used for the classification, perception, prediction, and creation of algorithms, which make it an important capability within the AI environment.

Comparable to Android in smartphones, TensorFlow is being used far more extensively than PyTorch, the competitive framework from Facebook. If nothing else, PyTorch provides a serious competitor to TensorFlow, a factor that will continue to spur innovation at both companies.

Alphabet currently trails Amazon and Microsoft in the cloud computing realm, but Google Cloud more than doubled revenues from $5.8 billion in 2018 to $13.1 billion in 2020.

Meanwhile, Alphabet's Waymo subsidiary is active in internet search, simultaneous voice, written translation, and autonomous driving. In gaming, Google's Stadia cloud service is gaining traction and the company continues to expand its library of online game titles. In consumer applications, the acquisition of Fitbit by Alphabet has increased the amount of consumer data that Alphabet can generate in what will be one of the largest markets for AI globally in 2030.

The health division of DeepMind was transferred to Google Health in 2019 to further empower Google's capabilities in the healthcare space. For example, in collaboration with the Moorfields Eye Hospital NHS Foundation Trust in India, the company developed an AI product to diagnose age-related macular degeneration, permitting the treatment and prevention of blindness in tens of thousands of people.[16]

Enter Facebook

In 2012, Mark Zuckerberg, Facebook's founder and CEO, realized that AI and deep learning was key to the future of the company. True, Facebook is currently building internet technology for immediate uses, not things like AGI that might not bear fruit anytime soon. But when Google acquired Hinton's services in 2012, Zuckerberg concluded that Google had made a critical move by getting to deep learning first. By the middle of 2013, Zuckerberg decided Facebook had to get there too.

A few days before the Google team met with DeepMind in London, Zuckerberg hired a renowned scientist with a lifelong passion for neural networks to head its new deep learning lab in New York City. His name was Yann LeCun.

Nearly forty years earlier, as a young engineering student in Paris, LeCun had stumbled upon a reference to neural networks in a book describing a debate between the psychologist Jean Piaget and the linguist Noam Chomsky. He was hooked, and the idea of a machine that could think like a human brain informed the rest of his career.

In the years that followed, LeCun explored neural networks, designed computer chips, and worked on self-driving cars. But his breakthrough contribution came in the 1980s. He took an idea he had first developed in Toronto—convolutional neural networks—and built a system at AT&T's Bell Labs in New Jersey that could recognize handwritten symbols with an accuracy unheard of at the time. The technology was sold to America's banks as a way of reading handwritten checks. At one point, Lecun's invention was clearing more than 10 percent of the checks passing through banks in the United States. Recognized as another pioneer in the field of neural network technologies, LeCun led a team of researchers and taught at New York University.

When Zuckerberg entered into serious discussions with LeCun about hiring him at Facebook, he described his vision for AI at the company: Interactions on the social network would be driven by technologies powerful enough to perform tasks on their own. In the short term, the technologies would identify faces in photos, recognize spoken commands, and translate between languages. In the longer term, intelligent bots would patrol the network, take instructions, and carry assignments, such as ordering a gift or making a hotel reservation.

LeCun agreed to join Facebook with a few key caveats: He would continue to work part-time at New York University, Facebook's new artificial intelligence lab would be built in New York, and Facebook's research would remain open-sourced. Zuckerberg agreed to all of LeCun's provisions.

Like Alphabet, AI and deep learning has profoundly changed Facebook. LeCun runs a fundamental research lab in New York that advances the technology's state-of-the-art capabilities. But much of the company's key AI work takes place in California with its applied machine learning team that focuses on implementing research in real-world products and services. The California team's goal is not to develop the most elegant, futuristic algorithms in its research. Instead, it has chosen to emphasize the practical application of AI technology and improve and grow services and ad revenues, according to Joaquin Candela, the Facebook executive who runs the California research team.

"You might be looking for the shiniest algorithm or people who are telling you they have the most advanced algorithm,"

Candela said in an interview with the Harvard Business Review,[17] "and you really should be looking for people who are most obsessed with getting any algorithm to do a job." For example, he said, rather than defining success as building the best algorithm, he defines success as deploying an algorithm that best helps users find a good restaurant in a neighborhood.

This two-tiered approach of investing in basic research while aggressively applying AI in the most practical ways played a major role in improving Facebook's services, boosting the number of users and ad revenues. AI and deep learning capabilities have been utilized throughout all of its core businesses, including Facebook (News Feed, Stories, Groups, Shops, Marketplace, etc.), Instagram, Messenger, WhatsApp, and Facebook Reality Labs with its virtual reality and augmented reality products s such as the Oculus platform and Portal.

When it first started focusing on AI in 2013, Facebook was interested in improving its ability to obtain value from its data. And to this day, the company's AI algorithms for classifying and analyzing data represent a key asset of the company. But the essence of Facebook is encouraging users to create data. The more users there are, the more data are created, and the stronger are the algorithms. Strong algorithms, in turn, help keep users online and target consumers for ad sales. At the end of 2020, the company's online social media and networking services had 2.8 billion monthly users who spoke 111 languages. Its investment in R&D in 2020 was $18.4 billion, with most of that devoted to software and AI-specific research.

But Facebook also became effective in using AI to recommend content, such as news and videos specific to each user. It also filters misinformation, hate speech, and violent content in posts, images, and videos. In addition, Facebook is developing a wide range of other AI-based capabilities, including computer vision, conversational AI, and natural language processing. For example, its AI can describe photos to the blind and visually impaired, provide predictive text suggestions while the user is typing, and more. Additionally, Facebook works with other companies on AI-related activities such as developing faster MRI scans and researching COVID-19 for drug discovery.

Transforming Microsoft

In 2014, Microsoft CEO Satya Nadella announced a cloud-first AI strategy for the company. It was part of a plan to revitalize Microsoft with AI and prevent the kind of missed opportunity as it had experienced in its mobile business. Nadella wanted to transform an outdated single software license business into an AI powerhouse.

As Nadella conceived it, Microsoft's mission was to bring AI to every large company, every one of their business processes, and every one of their workers. This made sense because Microsoft was already an enterprise-facing company, serving 90 percent of the Fortune 500 corporations with its familiar and trusted software. By contrast, Google, Facebook, and Apple have historically been consumer facing. As the market leader for operating systems and productivity tools in the professional and enterprise segment, Nadella knew that it was inevitable that these large companies would need to integrate AI and machine learning to stay competitive.

As a result of these changes in strategy, the company's market capitalization rose nearly 400 percent over the next six years. Today, Microsoft's own cloud platform, Azure, is the second-largest cloud provider, which contributed nearly a third of total sales in 2019 with around $39 billion.

The company placed the Azure platform at the center of its AI strategy. Azure has evolved from being a platform-as-a-service to being a cloud and edge platform that provides many AI services to industries, including manufacturing, retail, government, financial services, automotive, and others. This is occurring as the cloud market grows at *double-digit rates* and is expected to do so for years. The main drivers are the hunger for storage and computation, the Internet of Things, and multimedia. They require the creation and analysis of huge amounts of data.

Meanwhile, the company has been smoothly integrating AI into existing products, such as MS Word and Excel, Outlook, search, and the tagging of video content. Increasingly, it is assisting users with routine tasks that are detected by personalized digital assistants.

In 2021, the company announced the nearly $19 billion acquisition of Nuance, a company that provides speech recognition and conversational AI services. Nuance is best known for its deep learning voice transcription service, which is very popular in the health care sector. The two companies had already been working closely together. Before the acquisition, Microsoft was already using Nuance's Dragon AI technology in its health care solution, transcribing virtual visits, taking notes, and integrating information into patient health records. With the acquisition of Nuance, Microsoft had full access to its technology and was able to take its new AI transcription power beyond health care into other industries.

"Beyond health care, Nuance provides AI expertise and customer engagement solutions across Interactive Voice Response (IVR), virtual assistants, and digital and biometric solutions to companies around the world across all industries," Microsoft said in its blog.[18]

A sprawling company, Microsoft has also found seemingly countless ways to integrate AI and neural networks into its products and services. For example, LinkedIn, a 2016 Microsoft acquisition, uses AI to match job candidates with hiring companies, detect fake and inappropriate user profiles, and remove abusive accounts. Microsoft's Seeing AI describes images in words and places that language into Word, Outlook, and PowerPoint. The company has a $21.9 billion contract from the United States Army to produce visual augmentation headset systems based on Microsoft's HoloLens mixed reality glasses. And the AI in the *Microsoft Flight Simulator* renders the entire experience in 3D from satellite data transmitted every seventy-two hours.

China and the AlphaGo Earthquake

Eric Schmidt's May 2017 talk to Chinese executives came during a trip he made for another purpose: Google executives and Demis Hassabis's team from DeepMind had come to Wuzhen, China to defeat the reigning world champion of Go, Ke Jie, with an AI machine. Go is an ancient game, far more complex than chess, requiring powerful thinking and subtle, nuanced strategies that many believed only a human being could master at its highest levels of play. Hundreds of millions of Chinese people play Go passionately, so the head-on confrontations between a human and a machine had enormous symbolic value. DeepMind's contender was AlphaGo, an AI machine specifically designed for Go, with neural networks that had been trained by playing the game with other computers tens of thousands of times.

Despite the faith that many in China had in the superiority of human intelligence, AlphaGo handily won the five-game competition.

This was a wake-up moment for the Chinese leadership. The central government had encouraged AI development for years. Private companies as well as local governments had put large amounts of money and talent into the technology with impressive results. But the effort was not a uniform one and, as the AlphaGo victory demonstrated, China was a laggard in development and implementation of the most cutting-edge AI. Going forward, AI advancement to make China a global leader in this technology would be a top national priority.

Baidu

Baidu, China's leading search engine company, was launched in 2000 by Robin Li, who received his master's degree in computer science from the State University of New York at Buffalo and spent two years as a staff engineer at Infoseek in the United States. Upon his return, he and his cofounder, Eric Xu, built Baidu into the biggest search engine company in China and took the company public in 2005.

For years, Baidu was a remarkable success story. But in the mid-2010s, as China's consumers began to use their phones for just about every search and transaction, Li stubbornly refused to move away from desktops and lost ground to competitors. So, Li pivoted and began to invest in AI.

In 2017, the company had over 44.5 percent of the mobile search market and 80.8 percent of the internet search market in China, giving Baidu access to colossal amounts of demographic data, which it strategically channeled into the big data segment. It changed its public positioning from "a leading search engine, knowledge and information centered Internet platform and AI company," to "a leading AI company with a strong Internet foundation."[19]

Baidu built three state-of-the-art Deep Learning research laboratories in Silicon Valley with its focus on image recognition, machine learning, robotics, human-computer interaction, 3D vision, and heterogeneous computing. Baidu also established the Beijing Big Data Lab. The laboratories were driven by the mission of developing AI technologies that would impact at least one hundred million people. Their infrastructure was equipped with a powerful combination of neural networks, large datasets, and high performance computing.

Baidu has made a heavy commitment to its AI-driven Apollo autonomous driving platform. The rate of adaptation of robotaxis in China is far faster and more broad-based than in the United States. As a result, the adoption of the Apollo platform is expected to be much more rapid. Baidu is supporting robotaxi activities in Beijing, Chengdu, Guangzhou, and other locations. The company also supports autonomous buses that are appearing in Chongqing. In addition, Baidu is promoting its autonomous technology capabilities outside of China and has a license to pilot its robotaxis in Sunnyvale, California. Although its development of autonomous technology is still in an early stage, it is likely that by 2025 it will be deployed on a mass scale.

Autonomous transportation requires powerful AI capabilities and generates a large amount of data, and data analytics is one of Baidu's areas of strength. Baidu's activities in autonomous driving may allow revenues to be generated from hardware. But Baidu could also generate significant revenues by emulating a strategy of Tesla—selling its self-driving software alone at a high price. Again, China offers promising terrain for these automotive software sales. Chinese automobile makers sell almost thirty million automobiles annually while Tesla in 2020 sold only 499,550 vehicles.

Supporting Baidu's cloud capabilities is Baidu Brain AI services and its PaddlePaddle deep learning framework. At a time when momentum is growing quickly in China for the building of new applications, Baidu says there are over 2.6 million software developers on its open AI platform, DuerOS, who are integrated into devices like smartphones, televisions, and home appliances.

Baidu's big data initiatives are not confined to business alone. Many of its applications have demonstrated an impact on sensitive social and economic issues like controlling poverty, diseases, and stampedes.

Baidu's aggressive investments in AI are clearly paying off. In a recent head-to-head contest in natural language processing—which requires ten to one hundred times as many parameters as image-based learning models—Baidu's Ernie (Enhanced Representation through Knowledge Integration) model beat Google's Bert (Bi-directional Encoder Representations from Transformers) model. If nothing else, the win offered a glimpse of Baidu's potential.

Over time, Baidu has emerged as one of the top cloud-based technology companies in the world. Its bold ambitious initiatives like Apollo and cutting-edge technology platforms like DuerOS offer Baidu a major opportunity to lead in AI both domestically and abroad.

Cloud Capabilities of Alibaba Cloud, Google Cloud, Amazon Web Services, and Microsoft Azure

	Alibaba Cloud	Google Cloud	AWS	Azure
Cloud locations				
Regions	25	29	26	42
Availability zones	80	88	84	81
Services				
AI & machine learning	✓	✓	✓	✓
Analytics	✓	✓	✓	✓
Application integration		✓	✓	✓
Blockchain	✓		✓	
Business applications	✓	✓	✓	
Compute	✓	✓	✓	✓
Containers	✓	✓	✓	✓
Database	✓	✓	✓	✓
Developer tools	✓	✓	✓	✓
Gaming		✓	✓	
Healthcare & life sciences		✓	✓	✓
Hybrid cloud	✓	✓	✓	✓
Multicloud		✓	✓	✓
IoT	✓	✓	✓	✓
Management & governance	✓	✓	✓	✓
Media	✓	✓	✓	✓
Migration & transfer	✓	✓	✓	✓
Networking	✓	✓	✓	✓
Security, identify, & compliance	✓	✓	✓	✓
Storage	✓	✓	✓	✓
Virtual desktop				✓
VR/AR/MR			✓	✓

Source: International Business Strategies, Inc.

This table illustrates the cloud capabilities of China's Alibaba and the three U.S. companies: Google, Amazon, and Microsoft. China is competitive with the U.S. in data center technology today. With government funding, China is expected to have a larger data center capacity than the U.S. in 2025 to 2030.

Alibaba

In early 1999, Jack Ma, a former English teacher who had first encountered the internet on a trip to the United States, gathered eighteen people together in his apartment in Hangzhou. They had come to hear an idea he had for an ecommerce business that would use the internet to connect small- and medium-sized Chinese manufacturers with potential buyers around the world. After making a highly convincing pitch, he raised $60,000. Alibaba was born.

In 2004, after several years of growth, Alibaba made the fateful decision to take on eBay in China. Ma's company created Taobao ("Digging for Treasure"), a consumer-to-consumer website modeled on eBay that, unlike its American rival, did not charge a transaction fee on sales. At that time, however, China did not have an online payment system—only 1 percent of consumers had credit cards. So Ma invented one. By establishing accounts at banks across the country and an electronic escrow system, Alibaba gave birth to a secure payment system that would dramatically change Chinese consumer habits and culture. Ma was on his way to building an empire that would reach every person with internet access in China.[20]

Today, Alibaba is an enormous, multi-tentacled corporation, no longer dependent on American technology. It touches personal finance, entertainment, food services, car manufacturing, healthcare, and other industries. Alibaba has also established a number of independently managed companies—including Ant Group with its Alipay payment system—that has spurred the independent development of expertise and entrepreneurial skills.

Alipay, along with Tencent's WeChat Pay, dominates the billions of financial transactions conducted every day in China.

Alipay and WeChat Pay are used for almost every conceivable type of consumer transaction, including money transfers between individuals. In addition to supporting the implementation and payment of transactions, Alibaba has set up an efficient distribution and delivery system for goods. The company's Singles' Day event in China, which is the largest online shopping event of the year anywhere in the world[21], has become bigger than Black Friday in the United States in terms of the volume and the value of goods sold. Another division of Alibaba is Alibaba Cloud, a counterpart to Amazon Web Services, which is developing new hardware accelerators to support various types of AI-based transactions.

Alibaba has developed its own operating system for Alibaba Cloud's Elastic Compute Service; additionally, Alibaba Cloud recently announced a commitment of over $28 billion to expand its cloud ecosystem over the next three years. Alibaba has also developed its own operating system software for edge devices. These devices support many new initiatives of the company including smart cities, smart homes, and smart health.

The company has surpassed American tech firms in a number of areas. One example is Alibaba's "New Retail" initiatives that have injected new, AI-empowered services into both online and offline shopping experiences. In 2017, Alibaba opened futuristic grocery stores offering payment through facial recognition and thirty-minute deliveries. It created online smart fitting rooms where consumers are able to try on RFID-embedded clothing that enables the company to recommended new fashions to customers on the basis of style. Merchants are able to adopt Alibaba's AI-empowered "Store Xiaomi," a virtual customer service robot designed to respond to both written and spoken queries and help users find products when provided with a text, voice description, or photo. Seven hundred robots at Cainiao, Alibaba's logistic arm, are processing orders to load, unload, and run with no human employees involved.

Alibaba launched the City Brain in Hangzhou in 2016. It uses data from traffic cameras in the city to coordinate road signals, manage traffic flow, and reduce traffic congestion. The system worked dramatically well; it now covers eleven major areas of city life, including transportation, urban government, cultural tourism, and health, and has been adopted by more than twenty other cities across China.

Alibaba's support of financial transactions in China have surpassed those of any company in the United States. The company has said its ambition is to eventually support ten million profitable businesses across the globe that serve two billion consumers.

So successful is Alibaba, in fact, that the Chinese government became upset with the company's growing influence and the financial practices of Alibaba and Ant. In April 2021, China's State Administration for Market Regulation levied a $2.8 billion fine against Alibaba, equal to 4 percent of the company's domestic annual sales. The Chinese regulator said its investigation, launched in December 2020, found that the company punished certain merchants who sold goods both on Alibaba and on rival platforms, a practice known as "er xuan yi"—literally, "choose one out of two."

Previously, the Chinese government called off Ant's planned initial public offering that had been on track to be the world's largest stock sale and subsequently ordered Ant to revamp its businesses. Ant was suddenly subject to a bevy of new regulations, including turning itself into a financial holding company overseen by the central bank. The designation would subject Ant to rules similar to those governing traditional banks.

Nonetheless, Alibaba is still a world leader in big data and AI, including the processing, monetizing, and generation of data. Many of its capabilities are likely to be widely adopted by countries that are within China's sphere of influence, including those participating in the Belt and Road Initiative. Based on its current trajectory, it is highly likely that Alibaba will continue its technological progress to match and surpass American tech giants in the years to come.

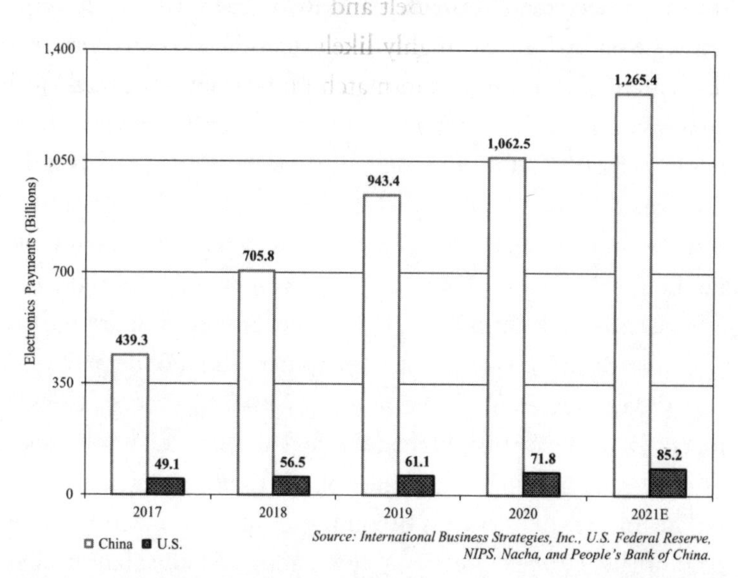

The number of electronics payments in China has accelerated rapidly. Relatively few Chinese people have ever had credit cards, making adoption of electronic payments easier than in the U.S. The Chinese government has limited the size of its currency notes to one hundred yuan (approximately $16.00), which makes it difficult to make large cash transactions.

Tencent

Like Alibaba, Tencent was founded by a man of modest background, roaring ambition, and a keen sense for the longshot opportunity. Pony Ma (no relation to Jack Ma) was born in South China's Guangdong province in 1971 and studied computer science at Shenzhen University. After working for a paging company, he set up Tencent in 1998. Tencent's first popular product was a free instant messaging program for computers called QQ, a copy of an Israeli program. It quickly attracted some investors, but by the early 2000s, the company was running out of cash. In their search for new sources of revenue, Ma and his team added services for a few cents each. One of them, which allowed mobile phone users to send QQ messages to computers, was suddenly a huge success. Eventually, QQ became China's primary means of communications on personal computers and mobile phones.[22]

Today, Tencent is, in the words of *Forbes Magazine*, "Google meets Disney, Netflix, Facebook, PayPal and Universal Music Group, plus gaming rolled into one." That description leaves out healthcare AI, which is becoming an increasingly important part of the Tencent portfolio. The company's subsidiaries offer payment systems, smartphones, entertainment, and most of all, WeChat, with its 1.2 billion users.[23]

Tencent has been tremendously successful in spreading the adoption of WeChat as a social platform and more. WeChat has been described as China's "app for everything" because of its wide range of functions, including text messaging, hold-to-talk voice messaging, broadcast (one-to-many) messaging, video conferencing, video games, sharing of photographs and videos, and location sharing. WeChat Pay competes with Alipay for digital transactions in Chinese restaurants, stores, and transportation among other environments. WeChat is an integral part of the daily lives of many Chinese, and its large database provides extensive information on buying patterns and other activities of the Chinese people.

In 2016, Tencent announced its vision of "Make AI Everywhere." It opened an AI lab in Shenzhen to focus research on machine learning, speech recognition, natural language processing, computer vision, and to develop practical AI applications for business in the areas of content, online games, social, and cloud services.

A top goal for Tencent is expanding its presence in healthcare AI. More than 38,000 medical institutions have a WeChat account and 60 percent of those institutions allow patients to book appointments online. Additionally, there are 2,000 hospitals that accept WeChat payment. These services allow Tencent to collect valuable consumer data that helps train AI algorithms. In a recent partnership with Babylon Health, WeChat users will have access to a virtual healthcare assistant. Pushing things even further, Tencent invested in iCarbonX, a company that aims to develop a digital representation of individuals to help perfect personalized medicine.

Tencent's internal R&D efforts resulted in the development of the Miying healthcare AI platform that helps physicians diagnose cancers and manage healthcare records.[24]

As Tencent moves into the healthcare space, many imaging and video technologies for gaming that Tencent controls can also be used for medical imaging. Medical imaging will require large storage capacity and high-processing throughput, which is similar to what is required for gaming. This may be a major source of revenues for Tencent in the future. Tencent is also expanding its entertainment company, with big data, virtual reality, and AI.

The government's recent restrictions on student gaming will put some limitations on Tencent, but its role in other areas will allow the company to grow.

The Future

Artificial intelligence is not a single monolithic block, but rather a cluster of different technologies with virtually limitless potential and countless future applications. Clearly, as AI capabilities are improved, new applications will emerge. American companies will continue to innovate and excel at some applications. Based on its current trajectory, however, China will be ahead of the United States in domains like internet AI, perception AI, and autonomous AI between 2030 and 2040.

Most of the American public, and even the nation's political leaders, are blind to the advances and potential of China's scientists, corporations, and cutting-edge technologies related to AI in China. This is because the growth of big data capabilities in China has been dramatic, but it is not as visible as the growth manufacturing in China and the resulting loss of American jobs. In the long-run, however, a Chinese lead in AI and related technologies will be far more consequential.

Chapter 3

Smart Warfare: The Emerging Digital Battlefield

"Whoever doesn't disrupt will be disrupted!"—Lieutenant General Liu Guozhi of China's People's Liberation Army[25]

In May 2018, Diane Greene, the head of Google's cloud computing business, announced to a gathering of Google employees that the company had decided not to renew its contract with the Pentagon for artificial intelligence work on Project Maven, an initiative that would help the armed services interpret video images recorded by drones. The program was deeply controversial at Google—and the news was greeted joyously by many of the 4,000 employees who had signed a petition demanding that Google refuse to do any work in the business of war. A week later, Google released a set of AI Principles that said Google would not create AI for weapons or other technologies that would cause injury to people.

For some officials at the Department of Defense, these announcements came like a kick in the teeth. Developing AI is unlike developing other military technologies. Fighter jets and submarines today are built by traditional, big defense contractors. However, the heart of innovation in AI and machine learning resides in the non-defense giants of Silicon Valley. Increasingly, in the years to come, AI will be essential for nearly everything the Defense Department does. To many in the Pentagon, the question was not whether AI would be used in new weapons systems. Like it or not, AI would be the core of future warfighting strategies. The question was this: On the battlefield of the future, whose AI technology would prevail?

What separated the pacifistic idealists of Google from the Pentagon was more than a difference of political opinion. It was a chasm of history, culture, and economics. That chasm limits the ability of the Pentagon to take full advantage of the vast intellectual resources of Silicon Valley in developing the next generation of defense technology. And it is one of many reasons why the use of AI in America's military lags so far behind the use of AI in America's commercial life.

For example, the most capable computer aboard a U.S. weapons system today is the core processor on the F-35 Joint Strike Fighter, an aircraft that took more than a decade to develop and which provides the premier air superiority and air strike platform for the U.S. Navy, Air Force, and Marine Corp. In the Pentagon it is often referred to as "The Flying Supercomputer." But its processing power is 800 times slower than a processor produced by NVIDIA that is available today on board a commercial car or truck. At home, F-35 pilots may be using advanced NVIDIA processors when playing video games or when they are driving their own cars off duty. But they are stuck with far slower technologies when at work.[26]

The distance that separates Silicon Valley from the Pentagon stands in dramatic contrast to the situation in China. There, the Chinese government can insist that a company or a scientist work for the defense establishment. In fact, the Chinese government is so close to its technology sector that it is sometimes hard to tell whether there is any boundary line at all separating the military from the nation's tech giants where work on AI is moving ahead at a fast clip.

But it was not always this way in America. It was the U.S. defense establishment that originally gave birth to a technology industry on the peninsula just south of San Francisco. It was defense contracts during and after World War II that turned a sleepy landscape of fruit orchards into a hub of electronics production and innovation now known as Silicon Valley.

In a memo written on April 30, 1946, Dwight D. Eisenhower told the War Department that America's weapons acquisition process had won the war and that it must be sustained in peacetime. He wrote that it was essential that major technology developments involve an integrated effort of civilian technologists and the military.[27]

"While some of our allies were compelled to throw up a wall of flesh and blood as their chief defense against the aggressors' onslaught," Eisenhower wrote, "we were able to use machines and technology to save lives."[28]

The rising threat of the Soviet Union in the late 1940s and early 1950s focused the public's mind and unified various segments of society. President Eisenhower facilitated this process by appointing what might be called military entrepreneurs, men with exceptional leadership characteristics who were granted all the financial and human resources needed to quickly achieve well-defined goals. These men included Air Force General

Bernard Schriever, who set up shop in an abandoned California church. He awarded giant contracts to private companies and integrated them into one military-industrial team. Five years later he had produced the Thor, Atlas, Titan, and Minuteman missiles that could deliver nuclear weapons to precise locations on the other side of the planet in minutes. Meanwhile, an aggressive, diminutive admiral named Hyman Rickover was able to miniaturize nuclear reactors and place them in submarines, enabling the vessels to run underwater almost indefinitely without refueling. And Edward Teller, a Hungarian physicist who had worked on the Manhattan Project, led a team that developed the hydrogen bomb in just three years.

In the 1950s and 1960s, a generation of engineers was drawn into the production of defense technology by the challenges facing the country and the interesting technical problems presented by Cold War competitions, such as the space race. There was money to be made, for sure, but keeping the nation safe from adversaries was considered by most people a patriotic and admirable thing to do.

Over time, the fast and entrepreneurial development of weapons systems was replaced by lengthy and bureaucratic processes managed by a burgeoning bureaucracy. Robert S. McNamara, the secretary of defense under Presidents John F. Kennedy and Lyndon B. Johnson in the 1960s, was the ultimate believer in the establishment of complex management systems in the interests of achieving efficiency. McNamara had worked out the systems and mathematics behind the firebombing of Japan late in World War II and had been president of the Ford Motor Company. Layers of management, oversight, and analysis were necessary, McNamara believed, to oversee a giant military-industrial complex. Congress became more intimately involved in the process of weapons acquisitions, inserting political priorities, such as insisting on the awarding of defense contracts to create jobs in certain congressional districts. These developments lengthened the time it took to plan and fund defense projects and made it more difficult for creative new ideas to be implemented. Eventually, the development time for major weapons systems stretched to a decade or more. Engineers and technologists faced the prospect of working on only a small number of big projects in an entire career.

In addition, the divisions over the Vietnam War that spread throughout American society in the late 1960s and early 1970s appeared among the engineers of Silicon Valley as well. Some engineers became disillusioned with working for the defense establishment and sought work on civilian projects.

Beyond that, there was a more subtle phenomenon. As *New York Times* reporter John Markoff wrote in his book, *What the Dormouse Said: How the Sixties Counterculture Shaped the Personal Computer Industry*, anti-authoritarian and anti-government values arrived in Silicon Valley in the 1970s. Those values mixed with the existing engineering mindset to create a unique and dynamic new Silicon Valley culture. It prized individual creativity, aggressive technological prowess, and a streak of libertarianism. At first, traditionalists in the Defense Department may have regarded this new generation in Silicon Valley as pampered hippies with stock options. But the fact was, they would build the most successful companies on earth.

As Silicon Valley grew and prospered, the defense industry consolidated. Driven by the fall of the Soviet Union and the ever-persistent drive for cost control and productivity, the number of major defense firms fell from 107 to 5.[29] More and more government money was spent tweaking existing platforms and weapons rather than developing entirely new ones.

By the turn of the twenty-first century, for many young technologists, the choice of working in Silicon Valley versus going to work in the defense establishment was an easy one: Silicon Valley could offer more money, more creative projects, and perhaps more fun. Moreover, Silicon Valley executives found the government procurement process calcified, slow, and difficult to master. To be sure, companies like NVIDIA and Microsoft did sign contracts with the Defense Department. But for many Silicon Valley senior executives, working with the government was not the most attractive option.

Until the 2010s, the Pentagon considered Silicon Valley a potentially useful partner, but not an essential one. That changed with the arrival of the new generation of neural networks and machine learning. The Pentagon realized that the integration of AI into the national defense would be essential for winning twenty-first century wars with major powers. The Pentagon now had an alarming vision of the future: a world of smart weapons, smart defense forces, and smart warfare in which the United States fell behind China in the development and use of AI. The Pentagon needed Silicon Valley.

AI and the Chinese Military

Ever since the victory of Google's AlphaGo in 2017, China's highest political leaders, including Xi Jinping, China's president and chairman of the Chinese Communist Party, have publicly placed the highest priority on the use of AI in the armed forces. As a result, the Party and the People's Liberation Army (PLA) are engaged in a full-court press to integrate AI into a multitude of military applications. To do so, the Chinese leaders have created a "military-civil fusion" of the private sector, academia, military institutes, and the PLA to advance "intelligentized" warfare by the armed forces.

The Chinese government has created an institutional infrastructure to oversee the process of military-civil fusion. Like in the U.S., the locus of the most advanced and creative innovation in AI in China is centered in the private sector. But, in contrast to the U.S., the Chinese seek to ensure that private sector progress in AI can be rapidly transferred for employment in the military. The process is directed by the Party's Military-Civil Fusion Development Commission, which was established in early 2017 under Xi Jinping's leadership. The PLA's CMC Science and Technology Commission exercises high-level leadership over the PLA's future research and development agenda and over military-civil fusion. It also serves a function intended roughly equivalent to America's Defense Advanced Research Projects Agency (DARPA).

In August 2017, the Ministry of Science and Technology and the CMC Science and Technology Commission jointly issued the Thirteenth Five-Year Science and Technology Military-Civil Fusion Development Special Plan to advance military-civil fusion and innovation. Under the Special Plan, the boundaries between civilian and military research and development are blurred. In addition, it assures that the PLA can work closely with cutting-edge AI researchers in the private sector. For example, a scientist named Deyi Li is president of the Chinese Association for Artificial Intelligence. He is affiliated with Tsinghua University and the Chinese Academy of Engineering. He is also a major general in the PLA.

Of its many military applications of AI, China has placed the greatest emphasis on unmanned systems. The Chinese military-industrial complex has used AI to develop unmanned combat systems, facilitate battlefield situational awareness and decision making, conduct multi-domain operations, and promote training. And the PLA is undergoing organizational reform and doctrine adaptation to determine how to operate AI-enabled platforms and effectively wage a new kind of warfare.

PLA strategists expect that autonomous combat by unmanned systems and the joint operations of unmanned and manned systems will dramatically disrupt traditional operational models. Leaders in the PLA and private industry anticipate that the future land, sea, air, and space battlefields all will be full of unmanned combat weapons, resulting in a "multidimensional, multi-domain unmanned combat weapons system of systems on the battlefield."[30] Among other things, the Chinese navy is experimenting with unmanned surface vessels and has developed unmanned submarines. The Chinese Rocket Force is leveraging remote sensing, targeting, and decision support for its missiles. And the PLA Army has focused on military robotics and unmanned ground vehicles.

Drones have been a particular area of focus for the military and the private companies working with it. The Rainbow family of high-altitude, long-endurance, unmanned aerial vehicles has been used for reconnaissance and strike missions. During the military parade on National Day in October 2019, China showed off its WZ-8, a high-altitude, super-speed, stealth reconnaissance, unmanned aerial vehicle; and Sharp Sword-11, a large, stealth strike drone.

China has also widely produced medium-altitude, long-endurance, unmanned aerial vehicles such as the Wing Loong platform (GJ-2) from the Aviation Industry Corporation of China and the CH-4, developed by the China Aerospace Science and Technology Corporation. They are less expensive than American drones and are actively being marketed for export. Within the PLA, the GJ-2 is used for integrated reconnaissance and precision strikes, including in joint support operations. And the Navy's ASN-216 can takeoff, land, fly, and film on reconnaissance missions without human supervision.[31]

The military has also set a high priority on swarm technologies that permit unmanned aerial vehicles to travel in groups or flocks with the intention of overwhelming enemies with sheer numbers. China's swarm tactics have roots in the massed ground attacks used by the PLA in the Korean War. In the spring of 2017, a formation of 1,000 unmanned aerial vehicles flew at the Guangzhou Airshow.[32] In another demonstration in 2017, 119 fixed-wing, unmanned vehicles made catapult-assisted takeoffs and demonstrated complex formations.[33] Future PLA swarms could be used for reconnaissance, strike, jamming, and other missions. Underscoring the nation's interest in this technology, China's Military Museum in Beijing includes an exhibit depicting a swarm combat system assault on an aircraft carrier.

China's military-industrial complex is also addressing what may be the greatest challenge posed by AI on the battlefield: AI generates an exponentially greater wave of data than earlier technologies, which decision makers must absorb, analyze, and act upon almost instantly. Confronted with more data than a human brain can digest quickly, the commander on the bridge of a warship or in a command center becomes a bottle neck for weapons systems and platforms.

A number of private companies and defense institutes in China are working to automate multi-sensor data fusion and information processing to enhance situational awareness on the battlefield. In addition, they are attempting to use deep learning to enable a substantial increase in the efficiency of intelligence analysis, including massive amounts of data. For instance, they are applying deep learning algorithms to the analysis of satellite imagery to greatly increase the speed of processing. The Huazhong University of Science and Technology's School of Artificial Intelligence and Automation, which has received support from the PLA's General Armament Department, has focused on intelligent information processing and pattern recognition. Among private companies, the AI start-up iFLYTEK is cooperating with the PLA on a voice recognition and synthesis module, which could be particularly useful for the PLA in intelligence processing. The PLA is also funding multiple other projects that address new forms of sensing and information fusion, including AI-enabled techniques for data and image processing and automated target recognition based on machine learning.

Another area of focus for China's military-industrial complex is war gaming, simulation, and training. PLA academics have often turned to simulations and computerized war gaming for training commanders in situational analysis and decision-making, particularly since the PLA has had little real battlefield experience in recent decades. This research includes extensive work on computerized war gaming for campaign- and strategic-level exercises. PLA researchers are focused on the potential for integrating AI into a military-simulations system to enhance the level of realism and create an artificially intelligent "Blue Force" to fight as China's adversary.[34] In addition, the introduction of AI within war gaming can enable its players to experience an approximation of future "intelligentized" combat.[35] In September 2017, the China Institute of Command and Control sponsored the first Artificial Intelligence and War-Gaming National Finals, convened at the National Defense University's Joint Operations College. It involved a "human-machine confrontation" between top teams and an AI system. The AI system was victorious over human teams by a score of 7 to 1.222.[36]

In China, the establishment of formal mechanisms and institutions to support the strategy of military-civil fusion enables effective synergies and deep collaboration between the PLA and the private sector. In a relatively short period of time, this strategy has produced a wide range of new military innovations. In the near future, the PLA can be expected to have a far greater capability to leverage private sector advances in AI than the U.S. Department of Defense has with Silicon Valley.

AI in the United States Military Today

The disruptions flowing out of Silicon Valley in the past fifteen years ambushed the defense establishment, and the Pentagon never really caught up. The Pentagon fell behind the commercial sector in AI and machine learning, cloud computing, edge computing, as well as in the centrality of data in the creation of algorithms for AI and machine learning.

Two more facts tell part of the story: Five of the largest American technology companies—Apple, Amazon, Alphabet, Facebook and Microsoft—spent approximately $70 billion on research and development in 2018. The five largest defense contractors spent about $7 billion in that year. In addition, those defense-contracting firms are largely hardware not software firms. They are not particularly adept at creating software. Often the weapons systems they produce are designed to have their software updated every several years. Compare that to Apple, which updates its iPhones and other devices every few weeks, and Tesla, which updates software every week.

To be sure, the defense establishment has authored strategies and the Pentagon has invested significant amounts of money in the development of AI and machine learning. This is particularly true for autonomous vehicles. The U.S. Army is pursuing a robotic combat vehicle, an optionally-manned infantry fighting vehicle, network-enabled glasses for infantrymen, and artificial and virtual reality aids for training. The U.S. Navy is developing unmanned underwater vehicles and unmanned surface vessels. The U.S. Air Force is pursuing collaborative attack weapons, swarming technologies, and a variety of unmanned aerial vehicles. Among other programs, DARPA is engaged in a multi-year program costing more than $2 billion called the "AI Next" campaign to enable AI to partner with humans to solve difficult problems.

But successes with the implementation of AI and machine learning in the military have been uneven at best.

Most Americans take for granted the AI they use every day, whether it is to pick a book, select a song, or decide on route home from work. But even these simple applications have not yet entered into the daily work of most military service members. These service members often have to engage in laborious and repetitive tasks that would have been taken over by machines in the private sector years ago.

One reason the Department of Defense has such a hard time integrating AI and machine learning into the ranks is the way it deals with its own data. In the private sector, companies long ago realized that data is the fuel of the information age and is the prerequisite for an information revolution. Machine learning algorithms require vast amounts of data, and companies like

Google, Microsoft, and Amazon regard it as a major competitive advantage that they have been amassing and stockpiling data over the years. By contrast, the military tends to dispose of its data as if it were the valueless and inconvenient refuse generated by more important activities. And even when military leaders do recognize the value of data, the military services tend to add more people to process it manually rather than apply machine learning at scale.[37]

The Department of Defense is also behind the curve on networks, sensors, and computing: the three building blocks of the information revolution.

Many of the military's networks are, by civilian standards, out-of-date and incapable of connecting with ease to other platforms and systems, if they can connect at all.

Christian Brose, a former aide to Senator John McCain and staff director of the Senate Armed Services Committee, wrote in his book, *The Kill Chain: Defending America in the Future of High-Tech Warfare*, that the Pentagon's "networks often are built by companies whose core competition is bending metal into military platforms...as a result, military networks are like a medieval world of unpaved roads, handmade bridges and checkpoints that inhibit more than facilitate the flow of information."[38]

The same is true of the development and implementation for military use of sensors. The Pentagon has spent billions of dollars on improving sensors with some impressive results. But so has the private sector. Civilian vehicles today are equipped with sensors that tell the driver what is going on around them at all times. Military vehicles do not. In addition, civilian applications made by companies like Ring and Nest allow consumers to keep visual track on their smartphones of what is happening in their homes or other important places. But at military bases, security is often maintained by troops stationed at key check points and patrolling perimeters.

The defense department also missed the revolution brought on by cloud computing. In the civilian world, the vast and growing wave of new data required a continuing demand for ever greater computing power. With the advent of the cloud, corporate consumers no longer had to purchase ever more powerful computers and create massive storage spaces in their offices. With cloud services, anyone could have virtually unlimited computer power and storage as a service. But it was only in October 2019 that the Defense Department finally awarded a contract to set up an enterprise cloud. That process was complicated by attacks by President Donald Trump on one of the competitors, Amazon, and its founder, Jeff Bezos.

The Pentagon has yet to catch up on the next generation of computing: computers on the edge. Until edge computing was developed, the handling of more and more data meant the building of more centralized storage. Large numbers of computer processors often had to be stacked together to handle the load, sometimes filling rooms or large buildings. With edge computing, the computing process is pushed out to the edge of the network: to the cars, homes, and appliances in the network that can collect, process, and communicate the information—the so-called Internet of Things. This ever-expanding network of smart devices could be used to create a more resilient and secure battlefield network. But edge computing has, for the most part, not been adopted by the U.S. armed forces.

Then there is the issue of the cultural resistance in certain parts of the military. Take the navy's submarine and aviation forces. Submariners have enthusiastically embraced AI-enabled, uninhabited undersea vehicles. Even during times of budget restrictions, the navy continued their development, building prototypes of increasing size over time, now including the new extra-large Orca. In fiscal year 2020, the navy's plan included $359 million for unmanned underwater vehicles with the intent to field 191 of various sizes by 2024. The submarine community has been innovating and experimenting with unmanned vessels in a variety of missions, including mine clearing and reconnaissance. By contrast, the community of naval aviators have actually pushed progress backwards in recent years. It has overseen the implementation of unmanned technologies for support functions like intelligence, surveillance, and reconnaissance—but it has resisted progress in settings where unmanned vehicles might compete with human-inhabited aircraft.

AI-enabled and unmanned strike aircraft are essential to maintaining the aircraft carrier's relevance and usefulness in future high-threat environments. In the late 2000s, the navy was moving towards this goal. Its 2006 Quadrennial Defense Review called for the development of "an unmanned longer-range carrier-based aircraft capable of being air-refueled to provide the greater standoff capability, to expand payload and launch options, and to increase naval reach and persistence."[39]

This program was designed to create a long-range, stealthy uninhabited aircraft to be launched and landed on aircraft carriers. The program successfully completed its goal of conducting autonomous take-offs and landings on a carrier by 2013. But the navy slowly pulled back the program, first to create a drone for use in non-hostile airspace. Eventually, the navy's plans for a future carrier-based uninhabited aircraft evolved into an unmanned aircraft whose primary mission is refueling manned aircraft.

During this time, China invested heavily in long-range, anti-ship ballistic and cruise missiles that would push carriers to operate beyond the effective range of carrier-based aircraft.

What accounted for the navy's retreat from the development of an unmanned strike aircraft?

Paul Scharre, a scholar of military technologies, has written that while pilots would still command uninhabited carrier-based aircraft, they would do so remotely from stations onboard the carrier and added: "Even if the resulting manned-unmanned team of aircraft made for a more effective carrier air wing, it would significantly change the way in which naval aviators conduct military operations. 'Top Gun' would have been traded for a trailer."[40]

This is not to say that the naval aviation community consciously attempted to thwart progress to the detriment of America's defenses. To the contrary, naval aviators are highly patriotic, highly skilled warriors, willing to make virtually any sacrifice, including their lives, to protect their country. But history is littered with examples of navies that lost battles and wars—from the Spanish Armada onwards—because they truly believed that the technologies they were trained in were superior to unproven, upstart technologies.

Finally, a great hindrance to the U.S. military's adaption of AI, machine learning, and other cutting-edge technologies is the difficulties associated with the Defense Department's procurement processes. America's technology start-ups—in industries ranging from information to communications to biotechnology—are one of the most important sources of innovation in American society. But it is striking how few do business with the Pentagon. In recent decades, only two such companies—Palantir and SpaceX—achieved unicorn status in the defense sector. And in both cases, before achieving that status, the young companies had to sue the Defense Department to get what they thought was a fair hearing.[41]

China, the United States, and AI in Battle

Comparing American and Chinese military AI at any one time is a snapshot of a moving target that is particularly hard to bring into focus. This is, in part, due to the fact that much about new military technologies is classified. But also the reality is changing fast. And these technologies are mostly untested in the real-world heat of battle.

Moreover, China's military and that of the United States have divergent approaches to enabling their defenses with AI and machine learning. China's agenda of "intelligencizing" its armed forces is a large-scale, top-down effort (allowing for some grass-roots initiatives) that seeks to incorporate AI throughout its force structure. The U.S., by contrast, is following a mission-driven model that is directed to a defined set of objectives with immediate relevance. For the PLA, AI is seen as an effective means of compensating for persistent challenges of human talent. The U.S. has usually recognized the skills of its human workforce as an advantage. While this provides opportunities for talented service people to play key roles, it raises the question of whether the U.S. military services will have difficulty in creating an AI-ready culture. The PLA is focused on the use of AI to support command decision-making and centralizing control through AI. The Department of Defense has primarily emphasized human-machine teams with "humans in the loop" and has been less averse to distributed decision-making.[42]

China as a nation certainly faces a number of challenges in AI. The U.S. civilian sector has the advantages of incumbency: the science behind AI and the majority of the algorithms in AI programs are American, as are much of the software and chip technology. Applying algorithms to particular deep learning projects requires a great deal of expertise and experience. China, at least for now, suffers from a lack of that kind of experience. While it is graduating three times as many STEM (science, technology, engineering, and mathematics) students than the U.S. (and many technologies are being developed for mass manufacturing by Chinese graduates of American universities), it will be many years before the collective experience of Chinese scientists, technologists, and engineers have a comparable level of experience.

China also lacks the ability to design and manufacture graphics processing units (GPUs) that permit large numbers of calculations to be carried out simultaneously and lacks field-programmable gate arrays that are used to train computers. These factors place a drag on AI development in and for the military.

But a big advantage for China is one of scale. Achieving breakthroughs in AI, and particularly in deep learning, requires vast amounts of data. As previously mentioned, China's size, burgeoning economy, and the centralized role of the government provide data as a resource in greater abundance than the United States. In addition, China greatly outspends the U.S. on funding for scientific research and development. The Chinese spent $429 billion, or 2 percent of GDP, on research and development in 2018 and invested $56 billion in AI start-ups in 2017.[43] By contrast, the United States government spent $127 billion on R&D in 2018. Private industry in the United States spent more than $400 million more on R&D in 2018.

China is also funding private companies to develop AI processors. An example is Horizon Robotics that is designing high-performance processors for real-time processing of data.

While China might be as much as five years behind the United States in developing processors and GPUs in 2022, by 2030, China will have caught up or even passed the United States in this realm. In the meantime, China is able to buy processors and GPUs from American companies. In addition, China is not limited in access for the most powerful chips. They can be bought in the open market.

In addition to processors, communications technology is very important. China is already using quantum communications, a highly secure means of transmitting data, in the military and in government. China also has multiple programs in quantum computing, a cutting-edge technology, with key goals of applying it to the military and in the aerospace industry.

In this complex picture, one thing is clear: It is critical that the ability of the respective militaries of the United States and China tap into the innovations, talent, and experience of their private sectors and academic institutions. And China's greater ability to leverage their private sector through its national AI strategy of military-civil fusion will give it a distinct advantage in the years to come.

The United States may still have superior military capabilities today, but the rate of change favors China. According to the National Security Commission on Artificial Intelligence's March 2021 report, if current trends continue, the United States will not lead in the near future.

If China does achieve military superiority in AI, the implications would be vast for the United States and its allies. As the United States becomes technologically weaker in relative terms, the chances of a Chinese invasion and takeover of Taiwan become exponentially higher, and freedom of passage in the South China Sea is endangered.

Making the Future Work

What does the U.S. military need to do?

The capabilities provided by AI will be the key factors in determining military power for both offensive and defensive purposes in the next few years. There will be the need for very high computer processing capacity in the cloud as well as in edge devices, including drones, robots, and other machines, in order to be superior to an opponent. There will, however, be the need to establish impenetrable firewalls around various cloud-based data ecosystems, which will become the targets of cyber warfare because destroying databases can win battles and wars.

Surveillance capabilities will be important, which include the imaging and communications satellites that will support a comprehensive communications structure for the distributed databases. Satellites are visible and may be destroyed or disabled by lasers that are on the ground, unless they are shielded from laser attacks. Consequently, land-based, high-bandwidth communications infrastructures for both wired and wireless networks need to be formed as backup capabilities. Military satellites will also be developed to either destroy or disable enemy satellites as well as Earth-based targets.

The aerospace and defense industries need access to the latest AI technologies for hardware, including support for high-performance signal processing as well as high-bandwidth connectivity. Israel is the global leader in many areas of signal processing, which includes image analysis. Many of the latest software capabilities for warfare AI analytics come from Israel, indicating Israel's effective adoption of new generations of technology.

In 2020, the U.S. military and aerospace applications use processors and other semiconductor products that are ten or fifteen years old, which do not optimize the benefits of AI. New products will need to be designed for the U.S. military and aerospace companies in twelve to eighteen months and manufactured with the latest-generation technologies so that the latest generation of semiconductor technologies can be used. Consequently, the U.S. and its allies will need to come up with radically different approaches to developing new systems. One option is to develop system design capabilities that use powerful AI algorithms and utilize high-performance computing capabilities in the cloud for implementing new product designs.

Smartphone vendors like Apple and Qualcomm can design the most complex chips in twelve to eighteen months. Military companies such as Boeing and Lockheed Martin cannot design semiconductor products with the same level of complexity as Apple and Qualcomm, even though the budgets of these companies are large and the time frames for development are longer than in the private sector.

The U.S. will need to establish leadership in supply chain capabilities for hardware and software to ensure their supply chains are well protected. A top-down approach is required for the use of AI in military and aerospace systems and to establish the appropriate supply chains for new-generation products and systems because many existing military capabilities will become obsolete as a result of AI. It is worthwhile to remember that many military battles have been won or lost in the past as a result of superior or inferior supply chains.

Computing power will increase by 10,000 times between 2020 and 2040, which will dramatically strengthen the capabilities supported by AI. In 2020, the computing power that is already available from Google and others is great, but that level of computing power is not widely available in the military. With the massive increase in computing power from 2020 to 2040, a key requirement for the deep learning algorithms used for war will be in optimizing the creation of strategies for war.

The level of computing power should be greater in the military than in the commercial sector. The approach of Google in its data centers is to connect 1,024 TPU (Tensor Processing Unit) processors in a pod. Multiple pods can be connected to create very high computing power. The military should do the same. In addition, the military will need edge computing or distributed computing networks, in which multiple locations are connected. This will build in multiple levels of redundancy and greater security.

With advanced distributed computing power, it may be very difficult to disable the hardware because computer systems will continue operating when some of its components fail to function. It is also important to use powerful encryption algorithms to protect the data. The ability to penetrate the enemy's encryption systems will also be part of future AI battles. If successful, this will give the U.S. the ability to destroy the data ecosystem of the enemy.

While it is important to have access to powerful computers to win battles and wars, there will also be the need to have strategies that are based on the utilization of AI analytics. The use of AI to develop these strategies as well as monitor their effective implementation will be vital for outperforming enemies.

In the near future, hypersonic missiles that are armed with nuclear weapons will be difficult to block unless superior AI capabilities can determine the potential threat of war and incapacitate the algorithms of hypersonic missiles to disable them from launching. (China has recently launched such a missile.) Consequently, even though military hardware is an important capability against some enemies, it will be software expertise and AI that will be the most important capability against a sophisticated enemy.

Chapter 4

Smart Healthcare: Two Nations, Two Sets of Challenges

The promise of artificial intelligence in healthcare and medicine evokes grand dreams of improving the quality of life for all humanity. Although still in its early stages, scientists and entrepreneurs envision a day of nearly flawless diagnoses, quick and inexpensive discovery of drugs, and genomic-based individualized treatments for all manner of disease. Machine learning and natural language processing, they assert, will alter specialties from radiology to psychiatry, increasing their effectiveness or, perhaps, replacing human specialists entirely.

In addition, leaders in both the U.S. and China see the potential for AI to address their health systems' most profound flaws. In the U.S., AI might be able to improve access to care, reduce spiraling costs, and lessen the disparities of care between rich and poor. In China, the Communist party envisions AI solutions to address the shortage of physicians in rural areas and make quality care available throughout the country.

In both countries, if implemented correctly, the ability of AI to improve the quality and length of people's lives is staggering. Some experts predict that AI use in preventive care could extend lives as much as twenty years. Already we are seeing a beachhead established for AI in preventive healthcare with products such as Apple's smartwatch—which has the ability to predict the onset of serious conditions such as heart disease—and sensors that can monitor blood sugar without drawing blood. But overall, the implementation of AI in medicine is no easy matter. Without national leadership in America, technological, cultural, and political barriers could delay those benefits for years. China does not lack that strong national leadership.

The ability to implement broad changes in U.S. healthcare is diffused among many powerful, conflicted players in a convoluted system in which government plays only a partial role. As a result, systemic change is slow, incremental, and politically fraught. By contrast, China, with its centralized power and leadership, stands a far better chance than the U.S. of using AI to achieve many quick and dramatic changes over the next two decades.

Comparison of Covid-19 Cases in China vs. The United States 2020-2021

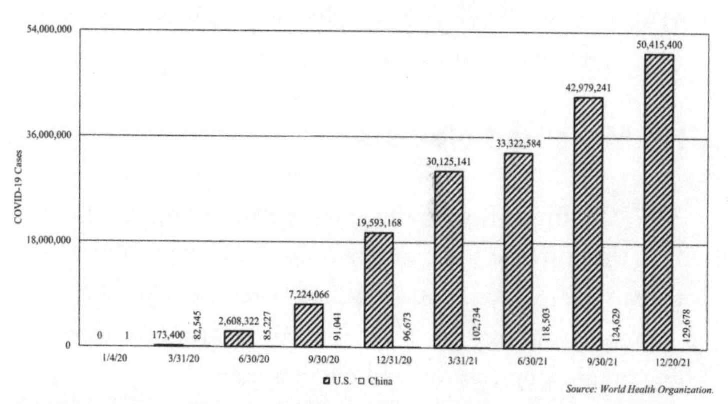

China's approach to controlling COVID-19 has been radically different from that of the United States. The number of COVID-19 cases in China has been much lower than in the United States, but at the cost of loss of freedom of movement for people in China.

The United States: Early in a Broad Revolution

Few Americans realize how AI, deep learning, and machine learning are seeping into the practice of medicine. But it is moving forward, with three themes emerging: One is that AI can improve the accuracy of many diagnoses over that of human medical professionals. Second, AI can spread the reach of specialists into fields and geographies where there are too few trained practitioners. Finally, it can drastically reduce costs. But the rise of AI in U.S. medicine is a complex process with implications that vary greatly by institution and specialty.

No More Radiologists?

In 2017, Geoffrey Hinton, the giant figure among AI scientists, said: "I think if you work as a radiologist, you are like Wile E. Coyote in the cartoon. You are already over the edge of the cliff, but you haven't looked down yet. There is no ground underneath. People should stop training radiologists now."[44]

While Hinton's prediction was premature, it may not be inaccurate.

Most radiologists read roughly 20,000 studies a year, or about fifty to one hundred cases per day. Each X-ray case involves at least a single image. Ultrasounds yield dozens of images. And CT scans and MRIs consist of hundreds of images. All together, there are more than 800 million medical scans a year in the U.S., or 60 billion images.

To cope with this volume of images, radiologists are trained to develop visual systems for identifying abnormalities quickly. In most cases, radiologists are able to perform accurate analyses. But these specialists, like all humans, suffer from tiredness, hunger, and biases. For example, radiologists can make mistakes because they suffer from "inattentional blindness," an inability to see something because they are so focused on seeing something else.

Some studies suggest radiologists' rate of inaccurate readings is far higher than previously believed, with false positives as high as 2 percent and false negatives as high as 25 percent. More than 31 percent of radiologists have experienced a malpractice claim, most of which were for inaccurate diagnoses.[45]

Meanwhile, machines empowered with deep learning have racked up an impressive record in reading radiological images. Numerous academic medical centers report the power of deep learning to sort through a wide variety of scans, including CT scans for liver, breast, and lung nodules. Scientists have used a deep learning algorithm trained to detect cancerous lung nodules and achieved an accuracy rate of 92 percent or better in a review of 43,000 chest X-rays. The findings compared favorably to board-certified radiologists. Geisinger Health in Pennsylvania used nearly 40,000 head CT scans to show high accuracy in the machine diagnosis of brain hemorrhages.

Progress in using deep learning to review radiological images is not limited to academia. Many American companies have entered the field. For example, Arterys received FDA (Food and Drug Administration) approval for its use of deep learning to review heart MRIs in 2017. Viz.ai received FDA approval for deep learning of head CT scans for stroke diagnosis the following year. And Zebra Medical Vision now uses convolutional neural networks to detect compression fractures of vertebrae with 93 percent accuracy. By contrast, radiologists rarely achieve better than 90 percent accuracy. Zebra Medical Vision, by the way, says its machines are capable of reading 260 million scans in twenty-four hours—a speed that would be 10,000 times faster than that of radiologists—at a cost of $1,000. No wonder that radiologists, who are often paid $400,000 a year, are worried about their future.

But not all experts agree that radiologists' days are numbered. Some predict radiologists will have a greater role in the new world of AI medicine, complementing machines that will do repetitive, more narrowly defined tasks. The arrival of AI, they say, might actually mean that radiologists will have the time to emerge from a darkened back room to meet with patients to explain test results.

"Radiologists can provide a more holistic assessment than machines can," writes Eric Topol, a world-renowned cardiologist and executive vice president of Scripps Research. "A narrow AI algorithm could prove to be exceptionally accurate for ruling out or pointing toward the diagnosis of lung cancer. But, in contrast, the radiologist not only scours the film for evidence of a lung nodule of lymph node enlargement but also looks for other abnormalities such as rib fractures, calcium deposits, heart enlargement, and fluid collection."[46]

Pathologists and Dermatologists

Much of what is said about AI and radiologists might be said of AI and some types of pathologists. They too are in the business of pattern recognition, a function that AI performs beautifully.

Pathologists come in a variety of specialties and subspecialties. The pathologists in question are those surgical or cytopathologic specialists who interpret slides of human tissue. Unfortunately, studies show that when more than one pathologist is asked to interpret a slide for cancer, transplant rejection, or other abnormalities, they often come up with highly different diagnoses. For example, in some forms of breast cancer, the agreement among pathologists can be as low as 48 percent.

Various studies show computer-driven diagnoses can do better. One Stanford University group used a machine learning algorithm to predict survival rates in patients with lung cancer, achieving greater accuracy than pathologists. Google used high-resolution images to detect metastasis with better than 92 percent accuracy, versus a 73 percent rate for pathologists, while reducing the false negative rate by 25 percent. Almost certainly, the eventual outcome in some subspecialties of pathology is a reduced role for humans. At the same time, many proponents of AI-empowered machines argue that these machines will take over the numbing, rote tasks and free humans to focus on higher-level analytical work.

Dermatology is another specialty where AI will have a major impact, solving a different kind of problem. Although 15 percent of all doctor visits in the U.S. involve skin conditions, there are only about 12,000 board-certified dermatologists in the U.S., a nation of 325 million people. The result is that most skin conditions are diagnosed by non-dermatologists and, by some estimates, as many as 50 percent of those diagnoses are wrong. In other words, if they can't do the job, there is a desperate need for AI-empowered machines in the field.

Every year, there are 5.4 million Americans with new cases of skin cancer. The key issue is to identify which cancers are the most common types of cancer—those with very high cure rates—and distinguish them from less common malignant melanoma, which kills about 10,000 Americans each year.

In an impressive study, a Stanford University team trained a convolutional neural network on nearly 130,000 images of skin lesions and biopsies of 1,942 lesions. The network out-performed twenty board-certified dermatologists in classifying cancer and, in particular, melanoma.

Because there are so few dermatologists in the U.S. and because dermatologists must also perform surgical procedures such as removing lesions, dermatologists are unlikely to be wholly replaced in the foreseeable future. What is more likely is that the vast majority of skin diagnoses that are now performed by non-dermatologists will be performed by machines. The result should be more accurate diagnoses for people in underserved regions and many lives saved.

Virtual Doctors of the Mind

In the United States, more than half the people who have psychiatric disorders do not receive care from a medical professional. There are fewer than eight psychiatrists for every 100,000 people in the nation. More than 106 million people live in areas that the federal government describes as short of mental health professionals. But virtual AI counselors are beginning to provide an answer.

In a number of recent psychological studies, participants were more comfortable talking with artificial counselors than with human ones.

One example is an experiment that took place at the Institute for Creative Technologies in Los Angeles. Half of the 239 participants were interviewed by an avatar named Ellie and were told that Ellie was not human. The other half of the participants were told Ellie was controlled by a human being. Ellie's questions grew quite personal, probing the participants' feelings. Their responses were turned into data used to quantify emotions such as fear and sadness, and also their openness to questions. By every measure, participants were more willing to disclose information when they were talking with a virtual human.

In another study, researchers at the University of Southern California developed software that was able to use seventy-four acoustic features to predict marital discord as well or better than therapists.

Such findings have spurred the private sector to action. For one, a company called Cognito has developed deep learning algorithms to monitor the mental health of patients through the way they speak. The technology picks up cues of depression and other mood changes. It has been used by insurance companies to handle incoming calls and by the Department of Veterans Affairs to monitor the mental health of at-risk veterans when they call a 24-7 phone line.

Other researchers and companies have found ways of gaining psychological insights through people's use of keyboards and Instagram photos with impressive results. In addition, companies, including Facebook, are now using AI to monitor voices and text in order to detect and prevent suicides. And there are numerous companies that provide mobile apps to simulate cognitive behavioral therapy (CBT), a technique of talk therapy that usually involves intensive human interaction. A meta-analysis of 3,400 patients using twenty-two smartphone apps for treating depression found significant improvement and a particularly high success rate for apps based on CBT.

Together, current trends in the United States indicate there could be a revolution in the way medical services are delivered for mental disorders, with dramatic increases in access to care for many underserved populations. In the near future, with the help of AI, patients will be availed of more accurate diagnoses based on a wide range of data on biological, physiological, anatomical, and environmental factors—factors way beyond the reach of a single therapist interviewing a patient in an office.

AI is being applied to countless other areas of medicine. It is likely that every specialty will eventually be touched. AI solutions will soon be able to empower pharmacists to give better advice about medications. AI generative adversarial networks can make dental crowns for individual patients more accurately than dentists. Other applications of AI are impacting the fields of surgery, ophthalmology, and oncology.[47]

But perhaps one of the most popular AI solutions among doctors involves electronic medical records (EMRs). The introduction of such records was regarded as a disaster; AI provides a solution. The EMRs were designed primarily for the convenience of insurance companies rather than physicians or patients. They require doctors to spend many additional hours entering data each day, leading to low job satisfaction and burnout. In addition, EMRs are the reason doctors conduct interviews with patients while typing and staring at their laptops, something most patients detest. But an alternative is provided by companies like Nuance—a pioneer in voice recognition and natural language understanding. Nuance offers physicians the technology to capture speech and transcribe it into EMRs so that a doctor can interview a patient hands-free, looking them in the eye. In the process, this technology saves doctors a great deal of time and effort. As mentioned earlier, Microsoft recently bought Nuance for more than $19 billion and expects to use its technologies in a wide range of other industries.

Drug Discovery

Drug discovery is a grueling, time-consuming, and extremely expensive process. Anything that would reduce the time it takes to develop a new drug—or reduce the chances that a new drug will fail—could save lives, save the developer large amounts of money, and potentially lower prices for consumers.

The goal of drug discovery is to identify small molecules that selectively modulate functions of target proteins. Without a lead or a hint, it can be a simple hit-or-miss process. Historically, guidance has been provided by naturally occurring compounds that have been used medicinally for hundreds or even thousands of years. In this way, the natural world has provided a large percentage of the molecules from which modern drugs have been developed.

In more modern times, many drugs that do not occur naturally have been discovered and developed de novo. And while drug discovery has historically been a haphazard process, modern pharmaceutical development employs strategies and methods of evaluating a very large number of molecules for specific desired targets. Yet even with modern methods of screening for targeted molecules, results are often failures because biological systems are complex, and molecules that work well for modulating one step in a process may have adverse interactions in other parts of the biological system. These issues may not become apparent until later in the drug development process, after a huge number of tests have been run, taking time and money but failing to return a viable result.

AI has been growing at a fast clip as a tool in the search for new drugs—even faster than it has for healthcare delivery. Since the mid-2010s, hundreds of private companies have been attempting to use a wide variety of approaches that include deep learning and machine learning to find drugs more quickly and effectively. In particular, machine-vision image analysis allows AI systems to predict which molecules might be effective for which biological targets, and thus accelerate the process of drug discovery. Similarly, simulations of chemical interactions can be performed to assess a drug's efficacy in disease treatment.

One of the great victories in the use of AI in medicine came in 2020 with the development of COVID-19 vaccines in less than a year. The process of such vaccine development had traditionally taken a decade or more. The fastest precedent for vaccine development was the mumps vaccine which took four years.

Pfizer and Moderna drew on digital technologies and artificial intelligence to roll out their vaccines. Together with the company BioNTech, Pfizer made it a particular priority to become more innovative and digitally focused. Before the pandemic struck, clinical staff members would visit trial sites in person. But during the COVID-19 initiative, Pfizer teams developed real-time predictive models of COVID-19 cases in specific countries, saving time and costs. Pfizer also created dashboards that used artificial intelligence to extract insights from vast amounts of data. For example, one dashboard helped researchers manage high volumes of information streaming from multiple external sources. Another dashboard created real-time predictive models for pandemic case rates in various countries.[48]

Other companies are using natural language processing to ingest all of what is known about drugs and molecules through biomedical literature and chemical databases. This has the advantage of allowing researchers to boil down vast amounts of data while avoiding the human biases that often inform such searches. For example, Insilico Medicine is working on cancer drug research that screens seventy-two million compounds from a public database to identify potentially therapeutic molecules that have not been patented.

Another approach is investigating the impact of drugs on human cells with the help of algorithms and automated microscopes. Recursion Pharmaceuticals, which performs image processing, has created virtual human cells in granular detail, including nucleus size and shape. The company then models thousands of molecules to see which could convert sick cells into healthy-looking ones. The company has identified more than fifteen new potential treatments with this strategy, including some that have moved forward into clinical trials.

Atomwise is another company involved in automated drug discovery, but it uses a somewhat different approach. Atomwise utilizes deep learning algorithms to screen millions of molecules. This process has led to nearly thirty drug discovery projects ranging from multiple sclerosis to Ebola. The company's AI technology, including 3D modeling, has provided more than seventy drug candidates with high probabilities of interacting with specific diseases.

Beyond drug discovery, American scientists are using AI to predict the proper dosage for new drugs. Deep learning algorithms are an excellent tool for such modeling because they easily incorporate patients' weight, age, gender, genomics, and the potential for drug interactions. This approach is being explored at the University of California, Los Angeles; University of California, San Francisco; and Virginia Tech.

AI at Hospitals and Health Systems

Hospitals are the biggest cost driver in the U.S. health system, accounting for one-third of America's $3.5 trillion healthcare bill. Many AI initiatives have taken aim at the underlying causes of the poor outcomes and rising healthcare costs at hospitals.

Avoiding unnecessary hospitalizations and avoiding readmissions are crucial issues for hospitals and health systems. Many AI studies have tried to predict whether particular hospitalized patients will need to be readmitted in the month following discharge, using AI to find features that are not captured by physicians. A study conducted at Mount Sinai Hospital in New York achieved 83 percent accuracy in predicting which patients would be readmitted by using electronic health records, medications, labs, and vital signs. The Mount Sinai team also studied electronic medical records from 1.3 million patients to predict diabetes, dementia, shingles, sickle cell anemia, and attention deficit disorder with a high degree of accuracy. Other health systems, such as Intermountain Healthcare in Utah and the University of Pittsburgh Medical Center, are now putting such predictive algorithms to work. In another initiative, Google is collaborating with health delivery networks to build prediction models from big data to warn clinicians of high-risk conditions, such as sepsis and heart failure.

There are also several firms that focus specifically on diagnosis and treatment recommendations for certain cancers based on their genetic profiles. Many cancers have a genetic basis, and human clinicians have found it increasingly difficult to evaluate all the genetic variants and their responses to a growing number of new drugs and protocols. Firms like Foundation Medicine and Flatiron Health, both now owned by Roche, specialize in this approach.[49]

Another important issue for hospitals is the rising need for palliative care and the lack of trained specialists to provide that care. There are only 6,600 board-certified palliative care specialists in the U.S.—or only one clinician for every 1,200 palliative care patients. As a result, less than half of the patients admitted to hospitals who require this care actually receive it. Meanwhile, of the Americans facing end-of-life care, 80 percent would prefer to die at home, but only a small fraction of them are able to do so.

The issue calls for becoming more efficient about care without compromising quality. To do so, a care team needs to know when a patient is nearing the end of life and when that person is likely to die. Human physicians are notoriously inaccurate in making such predictions. But AI-empowered machines can. A Stanford University computer science team developed a deep learning algorithm based on electronic health records to predict the timing of death. The team used a neural network to predict, with extraordinary accuracy, the times of death for a test population of 40,000 patients. More than 90 percent of people predicted to die in the following three to twelve months did so and more than 90 percent of the people predicted to live more than a year did so as well.

In another study, a Google team, together with three academic medical centers, trained a neural network with the records of 216,000 hospitalizations to predict whether a patient would die, the length of stay, unexpected hospital readmissions, and final discharge diagnoses. In addition, Google's DeepMind is working with the U.S. Department of Veterans Affairs to predict medical outcomes of more than 700,000 veterans.

The implications of these findings are of great significance for palliative care. They could also influence the use of resources by hospitals, including the use of intensive care units, resuscitation equipment, and ventilators.

The healthcare industry, the largest employer in the U.S., provides jobs to one in eight Americans and continues to grow faster than the U.S. economy. To put a lid on costs, many companies are thinking about how AI can automate operations and alleviate growth and related costs.

"It will soon be obvious that half our tasks can be done better at almost no cost by AI," says Kai-Fu Lee, a former Microsoft and Google executive who now runs his own venture capital firm. "This will be the fastest transition humankind has experienced, and we're not ready for it."[50]

Overall, administrative costs account for 20 percent of healthcare spending in the U.S. For example, hospitals and health systems employ more than 175,000 professional coders in America, at an average salary of $50,000, to review medical records and find the right billing codes for insurers. Paying for coders is a major part of every hospital bill, adding 15 percent to the charge for a doctors' visit and 25 percent for a visit to an emergency room. In addition, manual scheduling for operating rooms or staffing inpatient and outpatient units leads to extraordinary inefficiencies. And operators who make and cancel appointments—a task which could easily be handled by natural language processing—is another significant expense.

Many companies are working to increase AI engagement to upgrade workflow and efficiency in hospitals. One company, Qventus, uses data from billing systems, staffing lists, scheduling lists, electronic medical records, and nurse call lights to predict the activities in operating rooms, emergency departments, and pharmacies. MedStar Health, the biggest health system in the Washington D.C. region, uses AI to reduce the time it takes emergency room staff to review patient records, which can run an average of sixty pages. MedStar has developed a machine learning system that rapidly scans patient records and provides recommendations about the patient's symptoms, freeing doctors and nurses to render care. Hospitals and health systems are also using AI in robotic process automation for repetitive tasks like prior authorization, updating patient records, and billing.

Other companies are addressing inefficiencies in the operating room. For one thing, hospitals are seeing the growing appearance of surgical robots to enhance surgeons' abilities to create precise and minimally invasive incisions and stitch wounds. AI is also helping patients by performing some tasks that surgeons do poorly.

For example, Siddarth Satish, the founder of Gauss Surgical, a Silicon Valley start-up, realized in 2011 that the process surgeons use for estimating blood loss during an operation—guesses based on visual observations—was antiquated and highly inaccurate. Satish and his company were able to estimate blood loss with a high degree of accuracy by taking photos of blood in the operating room with an iPhone and using an AI algorithm to estimate blood loss. This technology lowered the rate of emergency surgical interventions overall and was particularly useful in delivery rooms, where blood loss is a major cause of maternal deaths. Gauss' solution has been adopted at more than fifty hospitals and health systems around the U.S.

Together, the progress in all these areas of American medicine is inspiring. But they are all in early stages. To make them a reality, the U.S. must overcome specifically American implementation barriers that could impede the realization of AI's progress for many years.

Healthcare and AI in China

Although China has undergone health care reforms for decades, high-quality resources are still concentrated in major cities and large hospitals. Smaller and rural hospitals and primary care settings suffer from a shortage of resources, including qualified doctors and other medical personnel. As a result, a large number of patients seek out experts far away from home. At rural and less popular facilities, both medical experts and patients are in short supply.

To overcome these challenges, China has made remarkable progress in integrating AI into the healthcare industry. The Chinese Communist Party and government have been pressing all healthcare and technology companies to introduce AI into the practice of medicine, including the tech giants Alibaba and Tencent. Among their primary goals: to build diagnostic tools that will make doctors more efficient and make quality care more available.

In all, more than 140 companies are applying AI in ways that could dramatically increase the efficiency of China's health system. In 2017, Alibaba's health unit introduced AI software that can help interpret CT scans and AI medical labs to help doctors make diagnoses. About a month later, Tencent unveiled Miying, a medical imaging program that helps doctors detect early signs of cancer, in the southwestern region of Guangxi. It is now used in hundreds of hospitals across China. Tencent is also very active in image diagnosis and drug discovery. It has invested in WeDoctor Group—a service that allows patients to video chat with doctors and fill their prescriptions online. VoxelCloud, an eye-imaging interpreting company also supported by Tencent, is broadly deploying diabetic retinopathy AI screening to counter the leading cause of blindness.

Other examples of the growth of AI in healthcare abound. A start-up called PereDoc, founded by a cancer radiologist, has installed its medical imaging AI algorithms in dozens of Chinese hospitals. Ant Financial, which has a chat bot that outperforms humans for customer satisfaction, has acquired the American company EyeVerify, which makes eye-recognition AI algorithms.

But the company that is considered by some as the AI leader in Chinese medicine is iFLYTEK, a major global player in speech recognition. In 2018, iFLYTEK launched an AI-empowered robot called Xiaoyi that passed China's medical licensing exam for human physicians with flying colors. Using the robot's ability to ingest and analyze a large amount of individual patient data, it plans to make its services available to general practitioners and oncologists across the nation.

iFLYTEK is also working with Guangzhou Hospital, which is using AI trained from 300 million medical records for almost every part of its operation. It organizes patient records, identifies patients through facial recognition, interprets CT scans, and manages operating room workflow. Using WeChat, people can get a pre-diagnosis online.

The public sector is also moving forward with AI to increase efficiency at hospitals and assist with the physician shortage. Xiangya Hospital of Central South University in Changsha, which serves an area of about one hundred million people, launched the first dermatology internet hospital licensed to operate online in 2018. Xiangya collaborated with over one hundred other hospitals across thirty provinces to establish the first national skin cancer collaborative network that enables faster access to diagnoses and subsequent treatment of cases. The hospital's skin care platform contains one million dermatopathology pictures and a tagged picture repository of 20,000 images. Based on those images, the dermatology department created a dataset containing 2,656 facial images showing six common skin diseases. Patients photograph their facial skin conditions with selfies on mobile devices and these are interpreted by an AI-based algorithm. The machine-driven results generally are as accurate as those of dermatologists.

Xiangya Hospital also deploys "smart cervical cancer screening" program, which assists in examining sixty million women annually by reducing the time it takes to evaluate pap test slices. It uses an interactive app that enables patients to upload their medical history, including pictures, ahead of an appointment, to free up time for the actual diagnosis when visiting a clinician.

America's Challenges

Despite the feverish activity in AI in the United States, the depth of American-based talent, and the capital available to invest in innovation, China may well be in a position to take the lead in healthcare AI over the long haul. That is because American healthcare resists systematic change, changes that challenge powerful interest groups, and changes that require government leadership. There are a number of reasons why:

The integration of AI into American healthcare is a difficult process. Aside from simply demonstrating superior efficacy, new technologies entering the medical field must also integrate with current practices, gain appropriate regulatory approval, and inspire medical staff and patients to invest in new ways of doing things.

For one thing, AI-based diagnoses and treatment recommendations are sometimes challenging to embed in clinical workflows and electronic health records. Such integration issues often are a greater barrier to the broad implementation of AI than any inability to provide accurate and effective diagnoses. In addition, many AI-based capabilities for diagnoses and treatment from tech companies are standalone. They address only a single aspect of care. It is hard for hospitals and health systems to use many standalone AI solutions that do not naturally integrate into each other or into systems already in place. Meanwhile, the major (and imperfect) electronic medical system companies have been slow to integrate AI into their software. As a result, it is up to individual providers, hospitals, and health systems to do the integration themselves, if they are willing to do it at all.

Errors, injuries, and subsequent litigation. Every medical institution is concerned about medical errors because they hurt patients. But in the U.S., the litigation system seeks to place blame for mistakes and threatens to place a heavy financial toll on those who a court decides are responsible for the errors. AI errors are potentially different from traditional medical mistakes for two reasons: First, patients and providers may react differently to injuries resulting from software error than from human error, particularly since an AI system does not necessarily explain how it reached a medical recommendation. Second, if AI systems become widespread, a problem might result in injuries, not just to an individual patient, but to hundreds or thousands of patients. Surveying this prospect, the legal departments of hospitals can be extremely wary of implementing AI systems.

Privacy concerns. Patients can be concerned about the large data sets that must be collected to make AI systems work. Hospitals that collect that data from many patients have been sued for sharing it with AI developers. In addition, AI can predict private information about patients, even though the AI system never received that information from the patients themselves. For example, voice recognition can signal depression, bipolar disorder, or other psychiatric diagnoses; a trembling hand on a keyboard might signal Parkinson's disease. Patients may consider it a violation of their privacy if the AI system's inference is shared with third parties such as banks or life insurance companies. Again, this could lead to litigation and punishing legal judgments.

Data availability. Creating an AI system requires large amounts of clean and organized data from sources such as electronic health records, insurance claims, pharmacy records, and consumer-generated information like purchasing information and fitness trackers. But that data is often fragmented across many systems. And patients typically see multiple providers and switch insurance companies. This leads to a splitting of data into numerous silos, systems, and formats. All of this increases the costs of implementing AI—when it is possible at all—and limits the entities that can adopt healthcare AI.

Cultural issues. As in the American military, there are frequently specialists whose intensive training and personal identity is invested in traditional ways of doing things. Like navy fighter pilots who resist the introduction of drones to aircraft carriers, some physicians honestly believe that transforming their fields with new technologies will neither benefit themselves or the public. For example, pathologists in the U.S. have resisted the introduction of the whole slide imaging digital technique, which has slowed the encroachment of AI into their field.

The politics of healthcare. As Paul Starr wrote in his classic book, *The Social Transformation of American Medicine: The Rise of a Sovereign Profession and the Making of a Vast Industry*, systemic change in U.S. healthcare requires the assent of myriad interest groups, including the pharmaceutical and biotech industries, the insurance industry, labor unions, medical device makers, physicians groups, leaders of the two political parties, and consumer groups such as the American Association of Retired Persons (AARP), among others. Starr points out that in cases of most initiatives proposing to systemically change healthcare in the U.S., a more powerful coalition of groups is opposed than in favor. Thus, political barriers have stymied attempts at change from Harry S. Truman's administration onwards. Even in the case of a successful initiative like the Barack Obama administration's Affordable Care Act ("Obamacare"), the legislation took nearly a year to wind its way through Congress and pass by a thin majority. It then faced a decade of political warfare and litigation. It should be noted that this legislation only brought change to a sliver of the American public, less than twenty million in a nation of 325 million. Although Obamacare addressed a serious problem—the uninsured—the changes it brought were tiny compared to changes that will come with the introduction of AI to healthcare.

In fact, AI can be expected to bring forth a new wave of issues that call for federal action. For example, it will probably take federal leadership to address the problems of assembling high-quality data sets across the nation in a manner consistent with protecting patient privacy. Despite spending $27 billion on incentive programs to encourage hospitals and providers to adopt electronic health records, there is no standard format or centralized repository of patient medical data. Without such uniformity of clean data, implementing AI solutions will be difficult. The government could provide infrastructural resources for data, ranging from setting standards for electronic health records to directly providing technical support for high-quality data-gathering efforts for hospitals and health systems that lack such resources. It is striking that 90 percent of U.S. hospitals have an AI or automation strategy in place up from 53 percent in 2019. But only 7 percent of hospitals' AI strategies are fully operational. Difficulties with data are one major barrier.

Just as importantly, the U.S. currently does not have a single federal-level law on data protection or security. Instead, the often overwhelmed Federal Trade Commission

has broad authority but little capability to protect consumers from unfair or deceptive data practices.[51]

Another area that needs the attention of Washington is how AI innovations are regulated and approved. The number of approved AI and machine learning medical devices has increased sharply since 2015, but there is still no specific regulatory pathway with the FDA or other agencies for gaining approval for such devices.

China's Advantages

By contrast, in China, the Communist party, government agencies, industrialists, and entrepreneurs work together without many of the barriers to AI implementation that America faces. While technological barriers abound, the Party and the government's top-down and centralized power structure gives China the ability to institute policies to change and reform institutions in ways that would be unimaginable in the U.S. The Party's ability to command obedience was demonstrated during the COVID-19 crisis in 2020, when it was able to test almost ten million people in Wuhan over a period of nineteen days. Issues such as litigation over patient injuries, privacy, and data sharing do not pose barriers to widespread implementation when those implementation initiatives have government support. And government support of AI development in healthcare has been intensive.

Monthly New COVID-19 Cases in China

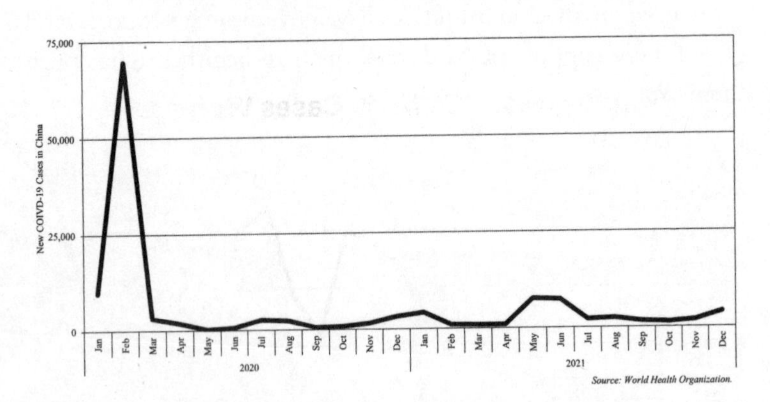

Source: World Health Organization.

Monthly New COVID-19 Cases in the U.S.

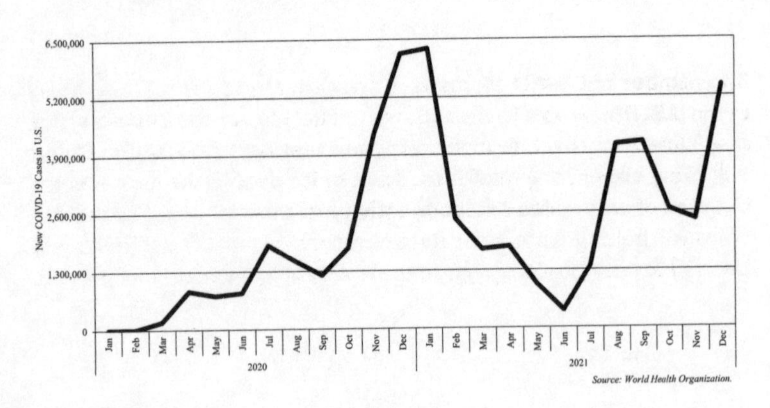

Source: World Health Organization.

Monthly New COVID-19 Cases Worldwide

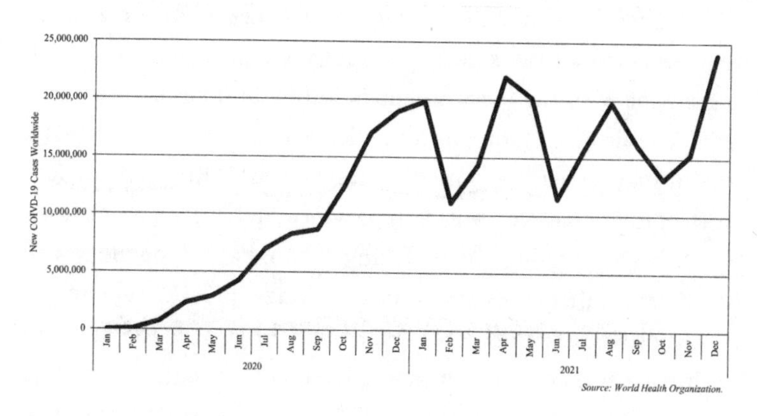

Source: World Health Organization.

The number of COVID-19 cases is dramatically lower in China than in the U.S. But people in the U.S. are unlikely to accept China's strict disciplined approach to monitoring and testing. In the future, China will likely collect the health markers of its people through a combination of wearables and connectivity to smartphones. These devices will be connected into data centers where AI capabilities will be used to provide visibility into the health of individuals and groups.

The Chinese government has been encouraging the application of big data and artificial intelligence in medicine since 2015, when the Ministry of Science and Technology and the National Health and Family Planning Commission launched a blueprint to support medical innovation during the Thirteenth Five-Year Plan (2016-2020). This involved guidelines on research in techniques in bioscience, precision medicine, and medical artificial intelligence. During two sessions in 2017, artificial intelligence was written into the China's *Government Work Report.* And in that year, iFLYTEK signed a strategic cooperation framework agreement with the government to jointly build a national artificial intelligence platform in medicine and implement artificial intelligence initiatives in medical research and clinical medicine. At the provincial level, nineteen provinces in China have released AI strategies, among them Beijing, Shanghai, Hangzhou, and Shenzhen. Other cities are in the early stages of AI development.

Another key difference between China and the U.S. in healthcare is data. The Chinese government has an unmatched resource in its ability to gather data (citizens cannot opt out of data collection). With the world's largest population that uses the same small number of mobile apps for transactions, China is awash with formatted data and has become a data powerhouse. In healthcare, China's electronic medical record system is used in nearly all hospitals as well as information from laboratory findings, imaging studies, follow-up results, and demographic information. Unlike in the U.S., that data as well as patients' financial information can be easily extracted.

In fact, Chinese leadership is increasingly regarding data of all kinds gathered by the private sector as a national asset that can be tapped as quickly as the state needs it. China's economic blueprint for the next five years, released in 2021, emphasized the need to strengthen the government's access to and control of data, the first time a five-year plan has done so.[52] Already it has begun pressing China's tech giants like Tencent and Alibaba to share data they collect from social media, ecommerce, and health-related businesses.

Beyond the vast quantities and centralization of data, gathering medical data is much easier for Chinese companies than for their U.S. counterparts: Chinese patient privacy protections are laxer and Chinese companies are not bound by as many rules and regulatory processes. The process of data gathering is also less expensive in China. Gathering a big trove of data in the United States typically requires negotiating with multiple partners who know the financial value of the data. IBM spent more than $3.5 billion between 2015 and 2018, amassing millions of patient records and billions of images used to build AI algorithms.[53]

Other factors work in favor of AI development in Chinese healthcare, such as government and venture fund investment, a supportive regulatory environment, and AI departments that now exist at most universities. But perhaps most important is that China's weaknesses in healthcare could become its strengths: its more sparse health system is easier to alter than America's advanced system with its complex structures and established interest groups. And China's leadership, always wary of social forces that could raise discontent, feels intense pressure to integrate AI into China's healthcare to address the gaps that could unsettle an increasingly affluent public.

Chapter 5

Smart Cars: The Three Revolutions

On January 28, 2021, General Motors (GM) made a groundbreaking announcement: America's largest automotive manufacturer pledged to phase out petroleum-powered cars and trucks by 2035 and sell only vehicles with zero tailpipe emissions.

At that time, it was a dramatic turnaround for a company whose main revenue streams depended on churning out gas-guzzling pickup trucks and SUVs. The American media lauded the company—once known as the heart of industrial America with a product that epitomized the American dream—for turning its back on the 130-year-old combustion engine.

That day, the headline of the story in the *New York Times* declared the move was "the most ambitious in the auto industry."[54]

What the *Times* story did not say was that the government of China had made the same pledge three months earlier, not for one company, but for all of China's auto industry. The article did not mention that GM and other western companies would be dependent on China for batteries and raw materials to build their electric cars. And they did not mention that the government of China, which had already invested an estimated $100 billion in transitioning away from polluting vehicles, had set a target of 2025 for widespread installation of electric vehicles on China's roads and highways and a target of 2035 for the fully autonomous delivery of people and goods.

While GM's announcement elicited surprise in Detroit, Chinese government officials could be forgiven if they considered the GM announcement no more consequential than a puff of exhaust smoke in a rearview mirror. Through government support and intervention in the marketplace, China's leaders had positioned their nation to dominate the world's auto industry in the years to come.

The automotive industry is undergoing three interrelated, AI-driven revolutions simultaneously: the rise of electric vehicles, the emergence of ride-hailing and shared mobility services, and the development of automated or autonomous vehicles.

Taken together, these three revolutions have the potential to fundamentally change the way people live on this planet. A new generation of vehicles using all three technologies could reduce traffic, pollution, global warming, highway deaths, and the war-torn politics of fossil fuel dependence. These vehicles could also lead to the shortening of commutes, the expansion of public spaces, and the reconfiguration of cities. Surely, it will generate trillions of dollars in profits for the innovators and manufacturers who do it best. Gaining or losing the dominant role in the three automotive revolutions could have epic implications for a nation's future wealth and power.

Beyond that, the automotive industry may provide the clearest test yet of clashing American and Chinese worldviews. The American belief that free markets, left to their own, are the most effective tool for industrial progress is butting up against

Beijing's approach: strategic industrial policies and investments to promote cutting-edge innovation and manufacturing.

The Car Conundrum

By almost any standard, vehicles powered by combustion engines are the most inefficient, mass-produced product on the planet. They are also environmental hogs.

For more than a century, the predominant mode of transport for Americans has been the personally owned, gasoline-powered, human-operated automobile. Most Americans believe that owning and maintaining their own vehicles is key to participating in contemporary society. But the cost is great.

More than 95 percent of the automobiles sold in the United States are gasoline powered. With the typical vehicle weighing about 3000 pounds and the average passenger weighing about 150 pounds, less than 5 percent of the gasoline energy generated is used to propel a passenger forward. And less than 30 percent is used to propel the car forward. The rest of the energy is wasted as heat and sound or is used to power accessories like air conditioners, radios, and headlights.

But the situation is even worse than that. In the U.S., where 85 percent of people travel by automobile, the average overall car occupancy is 1.7 people. On daily commutes, the average number of people in a car is 1.1, and that person travels in congested areas at an average of twelve miles per hour. Yet the cars, trucks, and SUVs that are most commonly sold hold at least enough room for five adults with engines so powerful that many can travel at 120 miles per hour or more.

These overbuilt cars are also dangerous because they are so heavy. The World Health Organization (WHO) estimates that auto crashes kill 1.3 million people globally each year and killed 42,060 in the U.S. in 2020. Auto accidents are the third most common preventable cause of death in the nation.

In addition, privately owned cars, on average, sit parked 95 percent of the time. That means owners have to figure out where to put them 95 percent of the time. A good chunk of your home is probably dedicated to the space for your car in the form of a garage. Then you must have a place to store it at work. And, of course, there must be a place for your car to park when you go to the shopping mall, the doctor's office, or the sports arena where your favorite team plays. In the years since World War II, vast swaths of American soil have been covered with asphalt and concrete to provide parking places for cars at rest.

Altogether in the U.S. today, there are 230 million licensed drivers who own more than 270 million light-duty vehicles, which burn more than 337 million gallons of gasoline a day. The emissions from cars and trucks constitute a fifth of the greenhouse gases created in the United States and therefore are a major driver of global warming.

Thus comes the dream of the three revolutions in cars: Electric cars with no polluting emissions; ride -sharing services like Uber and Lyft that can put cars to work nearly twenty-four hours a day; and automated cars that can dramatically reduce accidents since machines do not drink alcohol, do not get tired behind the wheel, and are not distracted by texts and children in the back seat.[55]

A study by the Lawrence Berkeley National Laboratory found that a fleet of automated and non-polluting electric vehicles operated by a ride-sharing service could reduce per-mile greenhouse gas emissions by between 63 and 82 percent compared to privately-owned hybrid vehicles.[56]

Another study conducted by the Paris-based International Transport Forum in 2016 looked at how shared mobility might change life in the future. The study examined the impact of replacing all car and bus traffic in Lisbon, Portugal, a mid-sized city, with fleets of shared automated taxis and shuttlebuses. It found that 97 percent fewer cars and buses would be needed to provide the level of transportation that currently existed, and 95 percent less space would be required for parking in the city.[57] In addition, a study by scientists at Columbia University estimated that with such vehicles, most American cities could get by with 15 percent of the current number of cars and buses and that the U.S. could reduce the national overall cost of mobility from $4.5 trillion to $600 million per year.[58]

In other words, the visionaries of the three revolutions believe automobiles and society may be on the verge of an unprecedented transformation. And key to the future of all three revolutions is AI.

America's Electric Vehicles

There is nothing new about electric cars. At the turn of the twentieth century, electric cars outnumbered gasoline-powered models. They had a number of advantages: they were quiet, easier to start, and easier to drive. The founder of General Motors, W. C. Durant, famously said at the time that gasoline-powered cars were "noisy and smelly, and frighten the horses."[59]

But electric cars did not improve as fast as gasoline-powered vehicles. By 1910, Henry Ford's Model T was selling for half the price. The electric car's major weakness—then and now—was the expense and weight of the battery. By 1915, less than 2 percent of the 3.5 million vehicles driving America's roads were run on electricity, and by 1935, production of electric cars had all but disappeared.

The next spark for serious electric car development came in the late twentieth century. That spark was California's air pollution. In 1990, the California Air Resource Board ordered car makers to start introducing zero-emission vehicles for 2 percent of their sales in 1998 and 10 percent of sales by 2003.

Not surprisingly, these measures met stiff resistance from the automobile and petroleum industries and were never achieved. But they spurred new research into electric motors and batteries for the first time in decades. GM produced a prototype known as the EV1 in 1996 and leased it to drivers who were impressed by its stylish design, speed, and power. But after producing 1,117 of the vehicles, GM decided to call them back and have them destroyed to save on development costs and avoid any potential liability.

Progress stalled once again until 2008 when a remarkable event occurred: the introduction of the Tesla Roadster, an electric-powered coupe with a sleek design that could outperform almost all gasoline-driven sports cars. Tesla captured the imagination of the media and the automobile industry. GM and Nissan followed suit in 2010 with the introduction of GM's hybrid Chevy Volt and Nissan's Leaf, a pure battery electric vehicle.

Tesla shook the car world once again in 2012 with its Model S sedan, an engineering triumph that *Consumer Reports* described as "the best scoring car we have ever reviewed."[60]

Elegant with a host of innovations, the Model S could carry up to five people 270 miles on one charge. While large enough to fit a family, it could outrace two-seat Porsche and Ferrari sports cars, rocketing from zero to sixty miles per hour in less than four seconds. Its innovations included a large flat screen that replaced all dials and knobs on the dashboard. Map data and algorithms updated automatically via the internet, something never done by an automaker. The major drawback was its price tag of $70,000.

The introduction of the Model S was a breakthrough moment. Within four years, car makers offered twenty-four EV (electric vehicle) models and had sold 400,000 electric cars in the U.S. In 2020, annual sales of electric cars in the U.S. totaled about 311,000, (down slightly from the previous year because of the COVID-19 pandemic) and was expected to more than double in 2021.[61] Overall, electric cars remained substantially more expensive than their gasoline-powered rivals.

One possible solution to the price differential may come from AI. The future success of electric vehicles depends on developing car batteries that are lighter, cheaper, more powerful, and more quickly rechargeable. Using AI, battery researchers can speed up the exploration and testing of a vast universe of new material formations that could make a battery. Discoveries that once would have taken years to achieve now are possible, in some cases, in a matter of weeks. The lithium ion batteries now in use make up 25 percent of electric cars' cost. AI research techniques offer hope that electric cars will reach price parity with gasoline-powered cars in less than five years.

An Electric Car Market with Chinese Characteristics

Across the Pacific, a very different tale from that of America's has unfolded as China speeds ahead of the United States in electric car production.

In 2020, China manufactured and sold 1.2 million electric passenger vehicles, more than the rest of the world combined. China had an electric car infrastructure that included a government-backed rollout of 800,000 public charging stations and an entire supply chain. For example, CATL, headquartered in Fujian province, was the world's largest manufacturer of batteries for electric vehicles. Chinese companies dominated the world's production of electric motors. China even had control of much of the world's production of the raw materials needed for electric cars, including cobalt, lithium, and rare earth metals.

The broad range of electric cars available in China is getting broader. The biggest seller in early 2021 was the Hong Guang Mini, a tiny vehicle with a maximum speed of about sixty miles per hour, a twelve-inch wheelbase, and the ability to carry four people in a tight squeeze. Its price was $4500. At the other end of the spectrum was the P7, a sleek, large, luxury model introduced by the start-up XPeng in 2020. According to the manufacturer, the P7 had a range of 400 miles per charge, a host of automated features, and a price tag lower than that of the Tesla Model 3.

How did China speed ahead of America in the electric car industry?

The answer is government support. For example, XPeng, the manufacturer of the P7, signed deals with investment funds linked to the city of Guangzhou, XPeng's hometown, and the surrounding province, Guangdong, worth $700 million. XPeng also received preferential terms on land, low-interest loans, tax breaks, and state subsidies that helped reduced the car's price tag.[62]

It was also the pollution for China that first sparked the electric car revolution. According to the World Health Organization (WHO), the death rate from air pollution in China is the highest in the world. Researchers at the Planck Institute in Germany estimate that contaminated air results in the premature deaths of 1.4 million people in China every year. Less than 2 percent of China's top 500 cities meet the WHO's air pollution standards, and cities like Beijing and Shanghai frequently issue red alerts for acute levels of smog and particulates. Factories and schools close, workers stay home, highways shut down, and flights are delayed.

The two major culprits are coal-burning plants and automobile emissions. With 300 million cars on the road, Chinese vehicle emissions standards lag Europe and the United States by several years. So the decision by the Communist party and the government to rapidly develop electric cars grew out of three goals: improving public health, politically assuaging a population who was increasingly disturbed by the air they breathed, and achieving global superiority in a cutting-edge technology. In 2010, China set out strategic plans to lead the world in electric vehicle production. By 2016, it had achieved that goal turning out approximately 500,000 electric and hybrid vehicles. As noted earlier, by 2020, China was producing 1.2 million electric cars with analysts projecting that they would continue to outpace the U.S. for the foreseeable future.

China has achieved its current electric car growth through government intervention in the market. Most of the 500,000 vehicles sold in 2016 were for city-owned taxi fleets and transportation companies or were purchased in Beijing or Shanghai, where government subsidies are especially large. But across China, subsidies for electric cars are also substantial, running as high as nearly $9,000 per vehicle. And in some cities, including Beijing and Shanghai, officials have waived quotas and licensing requirements to spur electric vehicle sales.[63]

The government support of electric vehicles in China raises key questions for the future of the industry: Are Chinese cars competitive in terms of quality? And can Chinese cars compete in markets outside of China? Certainly in the past, gasoline-powered Chinese cars were not known for quality and were not competitive in foreign markets. But Chinese automakers argue that their partnerships with foreign automakers such as Tesla and GM, the research laboratories they have established in California, as well as their growing base of experience, have given China the know-how to produce vehicles that are world class. They say their electric vehicles will soon make up for their current weaknesses in reputation and brand recognition internationally.

Vehicle Production: ICE Vehicles vs. EVs

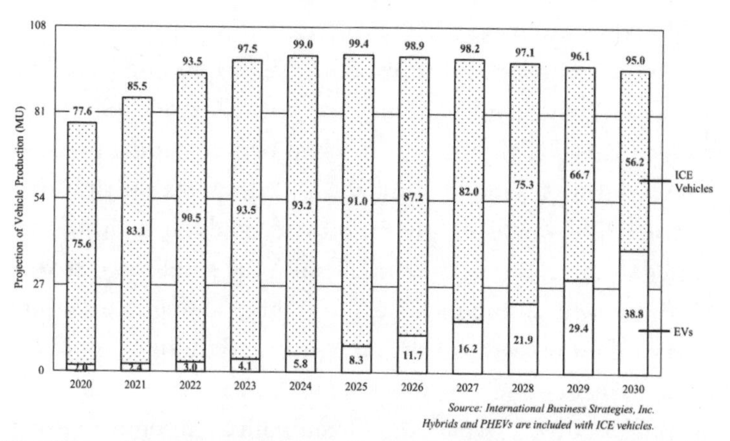

Source: International Business Strategies, Inc.
Hybrids and PHEVs are included with ICE vehicles.

The figure shows the rapid increase in volume of EVs and the decline in volume of internal combustion engine (ICE) vehicles. While the United States, Europe, China, Japan, and South Korea have committed to stopping production of ICE vehicles in 2035, other countries will continue to manufacture them.

Shared Ride Services

John Zimmer, who cofounded Lyft in 2012, once described his businesses as "the first phase of a movement to end car ownership and reclaim our cities." Lyft was, he said, a way to unlock unused cars, create economic opportunities, and reduce the cost of transportation.[64] While not immediately successful in achieving those goals, the business model was a direct assault on the idea of personally owned, gasoline-powered, human-operated automobiles that are so expensive to drivers and society.

For several years at the turn of the century, Zimmer and his partner Logan Green had a company called Zimride. Green had founded Zimride after he was inspired by ride-hailing practices he saw during a visit to Zimbabwe. Pedestrians on the street in Zimbabwe would simply raise their hands to indicate they wanted a ride and that they would split the gasoline costs with any driver who would pick them up. Functioning as an online ride board for colleges across the country and running minibuses in San Francisco and Los Angeles, the company never really took off. But in 2010, Zimmer and Green noticed an online upstart had been established in San Francisco called Uber.

Uber (first called Ubercab) was run by Garrett Camp and Travis Kalanick, a take-no-prisoners entrepreneur who had fewer social goals in mind. The idea was to use an iPhone app to provide a convenient way to hail a taxi in San Francisco, where taxis service is notoriously bad. The app allowed riders to summon a black limo with the tap of a finger and pay for the ride with a credit card. The service soon expanded to other cities and countries. In 2013, Uber introduced UberX, a more affordable service using the personal vehicles owned by the drivers, making it a true peer-to-peer platform. In 2010, the year the Uber app went on

Apple's app store, its business grew by 30 percent a month. In 2011, Uber expanded its services to New York City, and sixteen months after its founding it had revenues of more than $1 billion.

Zimmer and Green imitated the Uber business model by establishing Lyft. The two companies grew to become the Coke and Pepsi of the ride share world.

The difference between Uber and Lyft were more cultural than anything else. Kalanick was a notoriously prickly, domineering, and abrasive personality with an aggressive leadership style that eventually lost him control of the company. By contrast, Lyft's leadership was often criticized by investors for being too nice. Lyft at first began its rides with a fist bump between driver and rider, encouraged early drivers to stick a pink mustache on their front grill, and played up social aspects of the mutually beneficial form of transportation.

While Lyft and Uber started out with very different values, the two San Francisco-based companies converged on the same business model: taking a transaction fee from individuals driving their own vehicles. The two companies had few capital assets. They were mostly agile, computer companies that did not need to manage large inventories of equipment, employees, or facilities. Their major innovation was using computer algorithms that matched riders and drivers efficiently. Their apps removed the exchange of money from the rider-driver relationship, with fares automatically calculated and billed through the apps. They also balanced supply and demand by raising prices when demand exceeded supply. The companies established systems for passengers to rate drivers and for drivers to rate passengers, with drivers' ratings posted prominently on the app for riders to see and assess.

Artificial intelligence is central to the success of these ride-sharing enterprises. The most obvious benefit is using machine learning to identify the most efficient routes, so drivers can complete the assignment and get their payment more quickly without having to speed and endanger their passengers. But Uber's entire value proposition has artificial intelligence at its core. Each of its main product offerings heavily rely on machine learning. Uber's ride-sharing service uses AI when it comes to ETA predictions, customer support prioritization, one-click chat for drivers, destination prediction, and traffic forecasting, just to name a few. Uber Eats, the company's delivery service, uses machine learning to rank restaurants by user preference or estimate delivery times. AI is also used in fraud detection, risk assessment, safety processes, and marketing spend and allocation.

After their launches in 2010, Uber and Lyft were soon joined in the ride-sharing business by other mobility companies around the world, such as Chauffeur Privé in France, Ola Cabs in India, and Grab in Southeast Asia. The most successful long-distance, ride-sharing service is BlaBlaCar, a carpooling service. Founded in France in 2006 as a platform for carpooling, it transitioned in 2011 to a fee-based service. It connects drivers and passengers willing to travel together between cities and share the cost of the ride. By 2021, BlaBlaCar had more than ninety million members across twenty-two countries. BlaBlaCar says it doubles the occupancy rate of cars while operating a carbon-saving network. In total, the company says 1.6 million tons of CO_2 were saved by BlaBlaCar carpoolers in 2018, thanks to the relative efficiency of filled cars in comparison with alternative modes of transport.

Vehicle Ownership in the U.S. vs. China

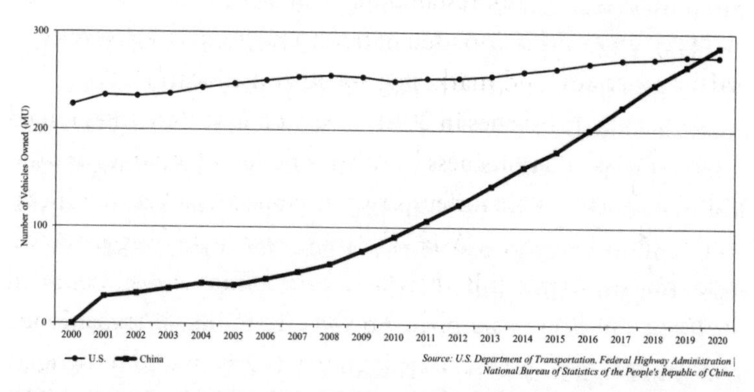

China manufactured approximately twenty-five million vehicles in 2021 and the number of vehicles owned in China exceeded the U.S. in 2020. China plans to address the congestion and pollution problems with the rapid adoption of EVs and fully autonomous transportation of goods and people by 2035.

Didi Chuxing: Flying Too Close to The Sun?

But it was not Uber, Lyft, or BlaBlaCar that became the great shared-ride business success story of the 2010s. It was Didi Chuxing, the Chinese ride-hailing company, which by 2020 had more than 550 million users in eight countries (Uber, by contrast, had ninety-three million that year) and more than 90 percent of the Chinese market. Didi, as the company is known, also rents cars on an hourly or daily basis, rents bicycles, operates minibuses, and delivers food and groceries.

Didi Chuxing was founded in 2012 by entrepreneurs Cheng Wei, Bo Zhang, and Wu Rui. Cheng previously worked at Alibaba while Bo worked at Baidu. In 2014, Jean Liu, a Harvard-trained former Goldman Sachs banker, came onboard and began leading a series of mergers and fundraisers that catapulted the company to become one of the largest and most talked about tech companies in China.

Liu's first task was to orchestrate a complex merger with Didi's main domestic rival, Kuaidi Dache, rebranding the new entity as Didi Kuaidi in 2015, and later as Didi Chuxing. Then came a larger challenge: repelling Uber, which had announced bold plans to conquer China after first entering the market in 2013. For two years, Uber's founder, Travis Kalanick, flooded the business with funds, subsidizing fares and drivers' salaries as part of breakneck expansion into dozens of cities.

Didi slashed its own prices in response. Both companies burned through cash—Uber lost $2 billion and Didi likely lost even more, but Didi was able to raise billions to replace it. As the war intensified, Liu struck deals, winning funds from Alibaba and Tencent. Not long afterwards, under pressure from its investors to stem losses, Uber quit. Liu brokered a peace accord, taking over Uber China in exchange for an 18 percent stake in Didi.

In May 2016, Liu unveiled a $1 billion cash infusion from Apple, part of a wider $4.5 billion fundraiser. By 2020, the company had raised $20 billion and had $12 billion in cash on hand. It lost money until 2021, when it reported a small profit in the first quarter.

As it expanded, so too did Didi's ambitions. Liu told *Wired* in 2018 that the company intended to become a global tech and transport giant by developing cutting-edge AI and helping revolutionize urban living in cities around the world. "We aspire to be a company that serves global transportation," Liu said.[65]

The company, she said, believes AI is central to creating new efficiencies and meeting China' growing appetite for rides. As *Wired* magazine noted, central to Didi's ability to manage its millions of drivers while learning the preferences of its passengers is machine learning. Didi executives are able to use AI to gather granular data on traffic across the nation. Pulling up real-time city maps on their laptops, they can monitor data from millions of cars in transit. They can see a map of Shanghai, for example, divided into thousands of hexagons. Each hexagon is just 500 yards across, glowing with red dots and showing traffic jams.[66]

In 2018, the company expanded to eight countries including Australia, Japan, and Brazil where it acquired ride-hailing app 99. In 2020, Japan-based Softbank invested $500 million in Didi's self-driving unit. Didi had been developing technology for its driverless cars for around three to five years, established a research center in California, and launched a pilot program for driverless cars in Shanghai.

Didi's success was unchallenged until its successful $4.4 billion IPO on the New York Stock Exchange in June 2021.

Suddenly over the following days, the Chinese government vented its wrath. It accused the Didi company of an array of misdeeds, including jeopardizing China's national security. The government ordered Didi to stop signing up new users, ordered its app removed from Chinese app stores, and sent investigators from multiple government agencies to occupy the company's offices. Analysts speculated that Didi, like many large companies in the Chinese tech sector, had become too powerful and independent and, as the repository of huge amounts of data, too likely to be subject to the oversight and control of foreign governments. In addition, successful entrepreneurs, like Liu, who is the daughter of the founder of Lenovo, incensed a public increasingly angry over income inequality and the wealth of tech business leaders. For the first time since it defeated Uber in the Chinese market, Didi's future was in doubt.

Autonomous Vehicles in the U.S.

Between 2013 and 2020, self-driving cars swept into the consciousness of the American public in wave after wave of hope and hype. In 2013, a report from Morgan Stanley said that sometime between 2018 and 2022, cars would be completely autonomous and that by 2026 they would dominate the marketplace. In 2015, Tesla announced that fully autonomous vehicles were just three years away. Ford said it would have fully autonomous cars by 2021. In 2016, the U.S. Secretary of Transportation, Anthony Foxx, declared that autonomous cars would be just about everywhere by 2021. In 2017, executives at GM, Volkswagen, Nissan, Google, and Intel were talking about the vehicles near-imminent arrival at scale. Investors poured tens of billions of dollars into the nascent self-driving industry and everyone from tiny start-ups to Google and Tesla, the big automakers, and Uber and Lyft were hard at work getting commercially viable autonomous vehicles onto the road.

Four years later, in 2021, little had actually changed in the way Americans drove their cars and trucks. The self-driving optimism of the preceding years had popped like an overinflated balloon. Uber and Lyft abandoned the pursuit. Three leading start-ups sold themselves to companies with deeper pockets. Only the richest players, like Alphabet's Waymo (formerly Google's self-driving car project), Tesla, and the giant automakers were continuing to plow billions into the endeavor. They spoke more modestly about arrival dates for commercially viable self-driving cars.

What had happened to the enthusiastic predictions?

For one thing, the leading players who came from the culture of Silicon Valley were used to moving fast, breaking things, and, when possible, apologizing for missteps rather than asking permission. That formula works well for the development of software. But producing automated vehicles is a different matter. In the highly regulated automotive world, moving too fast and breaking things can mean highway crashes, deaths, and serious legal jeopardy. For companies like Waymo and Tesla, things just took longer in the automotive realm than they had imagined.

Then there were the tough challenges faced by the engineers. They had to program vehicles to follow the rules of the road and communicate with human drivers and pedestrians. They also had to develop an array of sensors that could work flawlessly in all kinds of weather and visibility conditions. A vehicle would have to be able to figure out who goes first in a four-way intersection. It would have to comprehend that it should stop for a flock of ducks waddling across the road but should keep on driving when a cloud of leaves scatters in front of it, blown by a gust of wind. Cars with even the most powerful computers and the best AI might not be able to discern what to do at an accident scene when a police officer gestures to drive into a lane that is usually used by oncoming traffic.

And cars do not drive in a vacuum. There have to be changes in roads and infrastructure as well as federal, state, and local regulations to accommodate new fleets of self-driving vehicles. Not to mention the question of whether the public is prepared to ride in such vehicles. (Automated elevators at first ran into considerable public resistance in the twentieth century; who could trust an elevator to work correctly without a human to drive it?)

That is not to say that autonomous vehicles will never come to dominate the world's roads. In fact, there are plenty of reasons why they should.

The most clear-cut benefit is safety. Fully automated cars will be much safer than those with human drivers. Automobile fatality rates had been dropping in the U.S. and most industrialized countries for more than a century. But that changed in 2012, just as smartphones were becoming ubiquitous. Very simply, robots do not text. They also do not drink, eat, sleep, or discipline unruly children. Google estimates that automation could eliminate at last half of the 1.2 million crash deaths recorded each year around the globe.

Other benefits could also be large. Self-driving vehicles could greatly enhance access for young, old, ill, blind, and otherwise disabled travelers, as well as provide less costly transportation. If combined with electrification and shared riding, it would contribute to more livable cities.

The economic benefits also might be considerable. People would be able to use the time they currently spend driving to work, errands, school, and social encounters to do other things. Saving the time of a fifty-minute commute amounts to saving five forty-hour weeks each year. If you assume a worker earns ten dollars an hour, this would be equal to saving several thousand dollars for a worker each year.

Another benefit would be a reduction in parking spaces if vehicles, operated by companies, carry multiple riders. In Los Angeles County, for example, parking consumes 14 percent of all space, and the city has 3.3 spaces for every vehicle. Parking might no longer be needed at residences, along streets, and in city centers. It could be replaced with housing parks, bicycle lanes, cafes, and other public spaces. In addition, if automated vehicles made accidents rare or non-existent, vehicles could be smaller and lighter because they would no longer need heavy steel frames.

But the dream of fewer highway deaths and reconfigured cities assume that cars and trucks are fully automated or self-driving. According to the Society of Automotive Engineers, the automation of vehicles actually describes a spectrum of technologies, which they define in six levels. Level 0 describes cars with no automation. Levels 1 through 3 involve the increasing use of automation technology but require drivers to be attentive. By 2020, many car models used levels 2 and 3 technologies, such as lane keeping, adaptive cruise control, and automatic braking that enable drivers to take their hands off the wheel and eyes off the road, but only briefly. Level 4 cars are fully self-driving, only on certain routes and with a driver in the car. Level 5 is what most people have in mind when they use the term "self-driving vehicle." They will not need a steering wheel and no driver is necessary. Level 5 is, by far, the hardest to achieve.

A major controversy surrounds something that does exist today: cars with level 3 technology. Google and Ford have advocated skipping level 3 entirely, saying that this kind of partial automation—in which drivers can let go of the steering wheel at times but must keep aware of road conditions so that they can take control at a moment's notice—is unsafe.

Tesla, however, has equipped its cars with its Autopilot, a technology that approaches level 3. It has received national publicity about a growing number of traffic deaths involving the technology. At least three Tesla drivers died in crashes between 2016 and 2021 in which Autopilot was engaged and failed to detect obstacles in the road. In two instances, the system did not break for tractor trailers crossing highways. In June 2021, the National Highway Traffic Safety Administration released a list showing that at least ten people had been killed in eight accidents involving Autopilot and said it was conducting two dozen investigations into such crashes. Tesla and its chief executive officer, Elon Musk, denied there was a problem with the technology when used properly.

But critics charged the company had not made clear the limitations of Autopilot. Autopilot is not an autonomous driving system. It is most useful on major highways without intersections, but it is not designed to contend with the more complex challenges of urban traffic. While Autopilot is in control, drivers can relax but are not supposed to tune out. The car's driving manual tells drivers they are supposed to keep their hands on the steering wheel and eyes on the road, ready to take over in case the system becomes confused, fails to recognize objects, or a dangerous traffic scenario. But with little to do other than look straight ahead, some drivers seem unable to resist the temptation to let their attention wander.

Overall, what is the status of level 5 vehicles, the truly self-driving cars, in the United States?

The only real example of self-driving cars at work commercially in the U.S. in 2021, was Waymo One, a fully automated taxi service in the Phoenix area that had been in a testing phase for more than four years. Waymo, owned by Alphabet, said it planned to introduce the service in other regions of the country, and had stepped up testing in San Francisco in preparation for a launch there. Cruise, a start-up backed by GM, Honda, and several other investors, said it was aiming for a launch of fully driverless taxi services in San Francisco and Dubai by 2023. And Mobileye, an Israeli company owned by Intel, said it would launch fully driverless taxi services in a number of cities in 2023. But even self-driving car enthusiasts admitted that widespread adoption of driverless cars, despite the previous hype, was many years away.

Autonomous Vehicles in China

By most accounts, China is behind the U.S. in the development of autonomous car technologies and getting self-driving cars onto the nation's roads. But Chinese companies are rushing to catch up. The question is whether the technical challenges of self-driving cars and the unique traffic problems in China are too great to overcome—even for an entrepreneurial tech industry supported by aggressive government industrial policies.

Autonomous vehicles are a crucial element of the government's Made in China 2025 plan. In February 2020, eleven Chinese government departments jointly issued the Strategy for Innovation and Development of Intelligent Vehicles that sets forth a blueprint for how the Chinese government will promote the development of autonomous vehicles over the next thirty years. The Chinese government also announced a group of infrastructure investments that included a sixty-two-mile expressway connecting Beijing and neighboring Xiong'an that has several lanes dedicated to autonomous cars. It is to be built by driverless construction vehicles. And in January 2021, the Chinese Ministry of Industry and Information Technology released a draft policy permitting autonomous vehicle testing on highways. The short-term goal was to commercialize robotaxis, autonomous shuttles, and self-driving heavy trucks by 2025. The long-term goal is to sell and deploy Chinese autonomous vehicles abroad. The same is not true for American developers in China.

In 2021, Baidu announced the launch of what it called China's first paid autonomous vehicle service. By comparison with Alphabet's Waymo One, it was a modest affair. The initial service area was a little more than a square mile. By contrast, Waymo's service area in Phoenix was more than fifty square miles. But Baidu was also testing 200 fully driverless vehicles in three Chinese cities and testing its technology with safety drivers in two dozen more. The company's goals included deploying 3,000 robotaxis in thirty Chinese cities by the end of 2023. But its ambitions went beyond that. Apollo, Baidu's open-source autonomous vehicle platform, has 130 partners around the world ready to use Baidu's technology. Its Apollo Enterprise main product line includes solutions for autonomous highway driving, autonomous valet parking, fully autonomous minibuses, and an intelligent map data service platform.[67]

Apollo is already being used by some customers, including Udelv, an autonomous delivery van start-up that partnered with Walmart to test grocery deliveries. Today, with government support, Baidu's Apollo can support level 3 autonomous cars and will be able to support levels 4 and 5 by 2024.

Other Chinese companies were also moving aggressively. AutoX, backed by Alibaba, established a robotaxi pilot program in Shenzhen that was operating without in-vehicle backup drivers or remote operators. Didi's self-driving unit was operating forty automated vehicles in pilot programs and, after raising $800 million in investments, promised to expand quickly. In addition, Shanghai, home to the world's largest container port by volume, was working to speed up deployment of autonomous heavy trucks, and companies like Saic-iveco Hongyan and TuSimple were testing some fifty autonomous trucks.

Some aspects of China's unique environment could hinder the introduction of self-driving cars and trucks. The complex traffic snarls in its major cities will present significant challenges for even the most sophisticated AI systems. The country has highly complicated signage, with traffic lights and road signs not fully standardized. Also, right-of-way issues resulting from the frequent failure of Chinese drivers to follow the rules will not make it easy for autonomous vehicles.

But China also has unique market characteristics that could help make for faster adoption of automated ride-sharing vehicles than in the U.S. The Chinese are not as dedicated to the concept of automobile ownership as Americans, and China's rate of car ownership pales in comparison. WHO data shows there are 830 vehicles per one hundred people in the U.S., six times higher than in China. Additionally, Chinese cities are competing to get a head start on self-driving vehicles. Since the Chinese government greenlighted autonomous car testing on select public roads in March 2018, twenty-seven cities have awarded permits to over seventy companies operating around 600 autonomous vehicles.[68]

Most of all, the central government is doing just about all it can to support and protect China's autonomous vehicle development. Some government agencies forecast that China will spend up to $220 billion on 5G in the 2020s, including telecom networks that capture data from self-driving vehicles, cloud computing capacity to process the data, and mapping services to guide them. Mobile network operators like Huawei and China Mobile are building technology into their data systems to help autonomous vehicles on the road. Government agencies are also installing sensors on some roads to help autonomous vehicles move swiftly and safely. And the government is promulgating rules that limit the legal liability of self-driving vehicle makers, a situation that contrasts dramatically from that of the litigious U.S.

The Impact

China was never able to master the art of making and selling internal combustion engine cars. Now, with electric vehicles, it is aiming for world domination. With control of the supply chain for batteries, it controls a choke point for the most expensive and important component of electronic vehicles. And initial exports of XPeng vehicles suggest that Chinese vehicles can be competitive with Tesla on quality but sell for 30 percent less. If current trends continue, Chinese exports will have a devastating impact on legacy automobile companies, particularly in Japan and Germany where automobile production plays a central role in the economy. Even without Chinese competition, legacy car makers will have to drastically reduce their labor forces as they transition to electronic vehicles because their production requires only 30 to 40 percent of the labor involved in the manufacture of internal combustion vehicles. Already, Volkswagen has announced that the transition to electronic vehicles will result in the layoffs of 30,000 workers. The open question is how many of these legacy car makers in Japan and the West will survive in the long run.

Production of Electric Vehicles: China vs. Other Nations

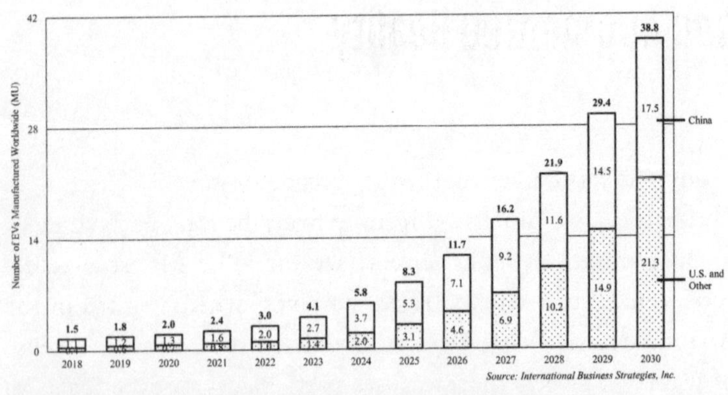

The production of electric vehicles (EVs) in China and other countries will increase rapidly in the coming years as the cost of EVs drops below that of internal combustion engines. China will lead the world in EV manufacturing, taking a significant share of the market from western car companies.

Chapter 6

Invented Worlds: Virtual Reality and Augmented Reality

Inside a cavernous virtual reality arcade on East 34th Street in Manhattan, one can virtually jump from the top of a skyscraper to the ground, virtually drive a race car at Le Mans, virtually cook a meal, or virtually battle "zombies" with ray guns in the Arizona desert. What attracts crowds to the arcade is a sense of reality so strong that it causes participants to experience an adrenalin rush that would be hard for any other amusement ride to match.

It is this sense of realistic and total immersion—what experts call "presence"—that gives virtual reality (VR) such extraordinary power and potential. It enables surgeons, soldiers, pilots, and firefighters, among others, to train at life-threatening tasks while not endangering anyone. It permits an architect to walk through and inspect a building that is still unbuilt. It can make all aspects of entertainment more compelling, from pinochle to porn. And it could permanently transform the way people interact with machines into a more organic process that involves bodily movement, sight, sound, feel and even smell.

Loosely defined, VR is a medium that uses imagery, sounds, and other sensations to make users experience what seems like a physical presence in another reality. Its sister technology, augmented reality (AR), allows for the viewing of real-life environments with digitally enhanced objects in it. AR does not create an artificial virtual environment. It alters the user's present reality by projecting data or images into the environment, only seen via the glasses or lenses the user wears. VR, AR, and mixtures of the two are often referred to as "extended reality."

VR today is typically experienced through the use of equipment, such as headsets with built-in speakers and display screens that serve as monitors. Global tech titans have produced popular VR headsets, which has resulted in strong market competition. Facebook offers the Oculus, Microsoft the HoloLens, and Taiwan's HTC sells the HTC VIVE. Both small and large developers have become successful in the production of software and content. Altogether, the global VR industry was estimated to be about $18 billion in 2021, with double-digit compounded growth predicted through 2026.

Many Years in the Making

The technological ancestors of today's VR and AR can be traced back to the flight simulators developed before World War II. The technology was totally different, but it was an attempt to create the reality of flying for a prospective pilot before he got into a real cockpit. But perhaps the first "device" that attempted to create a virtual world was the Sword of Damocles developed in the 1960s by a Utah professor named Ivan Sutherland. This was a complex and unwieldy set of goggles suspended from a ceiling. If you stepped up to the goggles and put them on you would see two primitive video screens displaying a transparent cube. If you moved your head, the scene changed to give you a different perspective. It did not actually do much, but Sutherland was later able to sell the technology for use in flight simulators.

In the 1970s, the U.S. Air Force developed flight helmets that could project useful information onto a pilot's field of vision. That project evolved into "Super Cockpit," a 1980s system of a flight suit, gloves, and helmet that would allow pilots to interact with 3D simulations of their flight instruments and see surrounding landscape on screens inside the helmet. (Later, the U.S. Air Force developed heads-up displays that projected data and images onto the front of the canopy of a fighter jet, in the pilot's line of sight. Today that technology is available in commercial automobiles.)

But the term "VR" did not come into use until the 1980s when Jaron Lanier, a cofounder of the small company VPL Research in Palo Alto, coined the term. VPL Research designed a set of goggles called the "EyePhone" that seemed to transport the wearer into a 3D reality. VPL also created a pair of gloves called "DataGlove" to go with the goggles and a "DataSuit" that let users see their hands, arms, and legs in a virtual environment. The problem: the three items together cost more than $350,000.

By the time VPL went bankrupt in the 1990s, the term VR had spread into popular culture. Hollywood flooded the market with films that included people with goggles traveling into immersive reality, such as the cyberpunk thriller *Johnny Mnemonic*, in which Keanu Reeves uses a VR headset and gloves to hack into a hotel in Beijing. Nintendo attempted to get in on the fad with *Virtual Boy*, a 3D video game that, unfortunately, tended to induce nausea. Nintendo pulled it from the market within six months.

By the end of the 1990s, the VR craze came to an end. The VR systems of the time had been far less impressive than the movies. They were extremely expensive, uncomfortable to wear, and rendered an experience that left users disappointed. Meanwhile, the internet had arrived, bringing the technologically inclined endless new wonders to explore, albeit not in 3D. Research continued in military and government laboratories. But the rest of the world had moved on.

VR devices fell from popular consciousness until 2013, when the world of techies became aware of the Oculus Rift, an extremely successful Kickstarter project that raised more than $2 million from people who wanted their own version of the Oculus headset. Reviewers said that the Oculus was finally producing an experience in which a player really felt like she was "inside" the game. Facebook bought Oculus for $2.3 billion in 2014. By the end of 2016, five different quality virtual reality systems were on the market for gamers and many more were under development.

Despite sales to tens of millions of consumers in the following four years, VR remained a niche product, unlike smartphones that sold billions of units. Companies creating VR hardware began to focus on the military and other businesses and industries that could afford higher-priced devices.

Facebook spent those years researching and developing ways to untether VR headsets from the PC (personal computer), freeing up a user's range of movement. It also worked on "inside-out tracking," a technology that monitors the position of the VR headset in relation to its environment, and wrote new algorithms that were more efficient and did not eat through a battery's power so quickly.

In 2020, with the arrival of the COVID-19 pandemic and people confined to their homes, sales of VR units in the U.S. shot up and Facebook offered the Oculus Quest 2 for $299, a product that did not require the use of a PC or cumbersome hardware. It sold more than five million of the devices in the first year.

In 2021, Facebook unveiled Horizon Workrooms, a virtual meeting room where people using VR headsets can gather as if they were at an in-person meeting. The participants could join with a customizable cartoon avatar of themselves. Interactive virtual white boards lined the walls so that people could write and draw things as if they were in a physical conference room.

Augmented Reality

AR's key trait is that it allows people to have access to virtual objects for comparison to real objects. For example, with AR glasses, an artist can view a virtual *Mona Lisa* as a reference while painting a rendition of the work and judge the level of similarity.

Gaming is projected to be the application that will stimulate the adoption of AR technology and create sales volume of tens of millions of devices. So far, however, it is industrial and commercial applications that are driving the adoption of AR, with companies like Microsoft and Google gaining leads in the field.

Niantic's *Pokémon GO* mobile phone app brought AR gaming to the mainstream in 2016. It allowed users to walk around the real world to catch, train, and battle virtual Pokémon characters. While *Pokémon GO* demonstrated the large potential demand for AR, the game's technical capabilities were extremely limited, and consumers quickly lost interest. *Pokémon GO* also demonstrated that smartphones weren't the right hardware for AR. What was needed was specialized smart glasses. Key requirements were that the glasses be light, easy to use, support voice and gesture commands, and have a long battery life. As of 2021, no such product was available in the consumer market.

But enterprise AR glasses, at a much higher price, do offer far more, including some models with voice recognition and hand gesture control. Microsoft's HoloLens 2, for example, also supports simultaneous voice translation, which will allow people who speak different languages to seamlessly communicate within the AR ecosystem.

While the volume of AR glasses in the enterprise market is smaller, companies like Microsoft and Google are gaining expertise they can apply to the consumer market in the future.

Extended Reality in Business and Industry

Gaming applications now make up slightly less than half of the VR industry. Increasing internet penetration, technological innovations, and advances in both gaming software and hardware are forecasted to drive the continued growth in that sector. But increasingly, VR is being used by the government and industries as diverse as retail, healthcare, aerospace, education, automobiles, and real estate.

Here are some of the fields where use of VR is underway:

The Military

The Defense Advanced Research Projects Agency (DARPA) and the various armed forces have engaged in massive research and development programs using VR and AR devices. Much of that investment has been in training and simulation. Teaching soldiers, sailors, and airmen to fight on simulators is far less dangerous and far less expensive than conducting real-world war games. For example, the $800,000 U.S. Army Stryker virtual trainer is more effective than real-world training exercises that can cost tens or hundreds of millions of dollars. The army has also developed a head-mounted display that projects data, information, and graphics into a soldier's view while he is also viewing the physical battlefield. In the future, these could also include bio-detection capabilities and could integrate with sensors that might be worn on clothing. In addition, VR simulators collect valuable data about how fighters react in various situations. Using AI, it is possible to spot patterns and trends that can lead to more effective combat tactics and strategies.

Civilian Training and Career Development

VR is also playing a growing role for training in the civilian world. In the U.S., companies spend more than $70 billion annually on training their workforces. Those expenses can be expected to grow as changes to technology and work accelerate. Online training has never delivered the realism of assembling a jet engine, mastering a sales technique, or installing a flow control valve in a nuclear reactor. VR can. With VR, it is possible for users to experience a fire in a high-rise building or a hostage situation before ever stepping into a real-world situation with the risks. With AR, users can see training manuals, specs, or data on the lenses that they wear. For example, Rolls-Royce, the manufacturer of jet engines, uses the technology to teach workers to assemble the critical components used in jet aircraft engines. A worker dons a VR headset and starts putting critical parts in place in an engine. If the worker makes a mistake, the system issues a warning and forces the technician to start over.

By the time the worker first gets to the real assembly line, he or she has learned the right sequence and placement for all the critical components.

Another possibility for training is the ability to hold conferences in virtual spaces and mixed-reality worlds. Accenture is developing "teleportation systems" that allow participants to view and walk through booths and displays at a virtual conference and view virtual representations of speakers and attendees, each represented by an avatar. Other mixed-reality environments have emerged as well.

Healthcare

Uses for VR and AR are emerging throughout the healthcare industry. At the University of Southern California, scientists are working with veterans to treat post-traumatic stress disorder (PTSD). Using a virtual space that resembles a computer game, the scientists build simulated environments where participants can reexperience their trauma in VR. The technique builds on a proven treatment approach called "prolonged exposure therapy" that allows them to confront and process difficult emotional memories. The approach is promising. A limited study conducted by the U.S. Office of Naval Research found that sixteen of twenty participants displayed significant reductions in PTSD symptoms after undergoing VR treatments. At Rush University Medical Center in Chicago, researchers have developed modules to help pharmacy and medical students recognize signs and symptoms of dementia. And at Virtual Reality Medical Center in San Diego, VR is used to help people with fear of flying, agoraphobia, claustrophobia, and fear of public speaking.

Biotech and pharmaceutical companies are also using VR for the design and development of drugs. In one example, AstraZeneca has developed a molecular visualization tool called Molecular Rift, which uses Oculus goggles. It creates virtual environments where a person can use gestures to interact with molecules and examine how they behave in different situations. Other uses of VR and AR in biotech include computer-aided engineering, safety studies, and air flow visualizations.

Extended reality is also being used in a range of applications by clinicians treating physical disorders. At the University of Maryland, physicians use AR to view ultrasounds while examining a patient. This makes it possible for a technician to avoid looking away from the patient during the examination. Physicians at George Washington University Hospital in Washington, DC, use Precision VR to view tumors, blood vessels, and other structures in the brain or other parts of the body. The system displays a patient's unique problem and helps doctors plan for surgery. It can also be used by the doctors to create personalized 3D virtual images to show the patient what will take place.

Engineering and Manufacturing

Designers, engineers, and architects were among the first professions to adopt VR and AR. Increasingly, they are using these technologies for visualizing spaces and prototyping things as diverse as buildings and cars. This allows them to see flaws in a design before anything is built to avoid cost overruns or safety hazards. With VR, an architect or contractor can step into a virtual space of a building that is not yet built to view technical specs, plumbing, or the design of an electrical system. Or, by wearing a pair of AR glasses during the construction phase, a builder can see data and graphics superimposed over an actual physical view of a room.

Boeing has used the Google Glass Enterprise Edition to streamline assembly processes in its factories. Aircraft assembly requires working with tens of thousands of components, including wires that come in many shapes and sizes. In recent years, assemblers have tried using PDF files to view assembly instructions. But using a keyboard while simultaneously assembling an airplane is a form of multitasking that turns out to be tedious, slow, and sometimes dangerous. The AR glasses used by Boeing delivers information in context that shows a worker how to find a specific wire, cut it, and install it. Workers see the detailed wiring instructions in the corner of their lenses. The app that drives the system incorporates voice commands and touch gestures. According to Boeing, the technology significantly reduces the assembly time and error rate.

In 2020, AR was used for the installation of complex semiconductor manufacturing equipment because engineers during the pandemic could not travel to manufacturing facilities. That began building up an experience base for semiconductor companies that saw the efficiencies they could realize through AR and remote engineering.

Agriculture

In agriculture, farmers can install high-resolution cameras around crop fields to send pictures and videos to a server or smartphone. Using AR and AI, a farmer can then compare the growth and appearance of actual crops to ideal crops so that the farmer can adjust irrigation, fertilizer, and other options.

The death of bees has stimulated scientists to develop new technologies to perform controlled pollination on fruit and nut trees. Drones, for example, can be programmed to autonomously fly a specific route and spray pollen on nut trees. Fruit farmers can then use AR capabilities to target the particular blooms for pollination in a flower cluster to produce the best fruits on a tree.

In the future, there will be the increasing use of AR to select fruits and vegetables that are ready to be shipped to consumers. When picking produce for delivery to consumers, a robot picker with AR capabilities will compare the actual vegetable to an ideal vegetable to see if it is ripe or rotten before it is put up for sale.

Education

Online education has opened up a series of possibilities for students from kindergarten to college. They are able to view and store data and information and proceed through a course at the learner's own speed. But as many school systems learned during the COVID-19 pandemic, the results of online learning can be disappointing; it does not deliver a particularly dynamic experience. At some point, human interaction is essential.

With virtual reality and AR, a student can experience a 3D school without traveling to a campus or setting foot in a classroom. Students can visit a zoo, view the signing of the Declaration of Independence, or walk on the surface of the moon. Using avatars, virtual representations of physical objects, and graphical interfaces, students can also experience a lecture in a lifelike way. Students can view everything from molecules to mathematical formulas in 3D shapes and forms that make sense. They also can interact with teachers and other students in a natural manner.

Research indicates that VR, AR, and mixtures of the two can enable learning effectively. A study of medical students at the University of Saskatchewan found that the use of VR improved learning accuracy by 20 percent. A study in 2018 at the University of Maryland showed that participants improved overall recall by 8.8 percent.

Marketing and Retail

Extended reality permits customers of retail organizations to have more authentic shopping experiences by moving through a commercial space in a more realistic way and viewing merchandise from various perspectives. In the future, with the use of haptics, a technology that stimulates the senses of touch and motion, and other feedback technology, customers will be able to feel the texture of a shirt or a sofa. Today, many stores are taking steps in that direction. Sephora, the cosmetics company, has turned to AR technology to allow customers to see what lipstick, eyeliner, or makeup looks like on their faces digitally. The app allows customers to view themselves with makeup on in real time. They can move their heads at different angles and the makeup follows. Sherwin-Williams's paint app allows prospective customers to see what various shades would look like in an office or a bedroom. And Lowe's, the home improvement chain, introduced a virtual realty shopping experience that allows customers to experience a home improvement tutorial and get hands-on experience attempting various projects. This initiative allowed the company to gather data on how customers reacted to the tutorials and where in the process they became confused or frustrated.

Experts in the real estate industry say that VR may revolutionize their business by permitting customers to walk around properties that are on sale without having to travel around town.

Auto manufacturers are also using VR technology. An app from Audi allows a prospective buyer to use a headset at home to view and explore the interiors of a vehicle in three dimensions. Audi also offers an app at its dealerships that allows customers to view different colors and options using Oculus or HTC headsets. And a third app permits prospective buyers to enter the virtual cabins of any of Audi's dozens of models and, using headphones, hear what the car sounds like.

Travel

The online travel industry is now a $700 billion business that allows people to book hotels, flights, cruises, and tours with a wealth of information and steep discounts. But the online experience fails to show people what a destination is really like. It also does little for people who, because of physical or financial limitations, are only able to travel vicariously. In a virtual space, it is possible to walk through the castles of Spain, a Hong Kong market, the glaciers of Iceland, or along the beaches of the Bahamas. With the help of VR, it is possible to stroll through a resort to determine if the rooms, the pools, and the restaurants are really where you want to go. VR provides for a more personalized and intuitive experience than a website or travel brochure ever could.

In 2015, Marriot Hotels introduced an in-room VR experience called VRoom Service that displayed VR postcards with immersive travel experiences. The next year, Lufthansa set up VR kiosks at an airport in Berlin. Passengers waiting for their flights could don an Oculus headset and take a virtual tour of Miami or the Great Barrier Reef. Later, the Smithsonian Institution, together with the Great Courses, introduced an immersive journey to Venice, Italy in which the visitor experiences a tour of the major sites with a historian in a gondola that rocks from side to side. A turn of the head revealed a 360-degree scene and the gondolier pushing at the back of the boat.

AR has a role in the travel industry as well. A tourist just has to point her smartphone at signs, menus, and other printed materials for Google Translate and other apps to provide instant translations. At Gatwick Airport in Great Britain, an app projects a path of green arrows to guide a passenger to the correct gate; the user simply points her phone and sees the arrows on her screen superimposed over the image of the real airport scene.

China's Growth in VR

Due to the size of China's domestic VR market and government support, China has become a key global player in the VR industry. The market in China exceeded $8.5 billion in 2020, making it the largest VR market globally. In that same year, the country accounted for 54.7 percent of commercial and consumer spending on VR and AR worldwide and is growing rapidly, according to the International Data Corporation.

The Chinese government has set a goal of 2025 to achieve worldwide leadership in the extended reality industries. It has also taken a leading role in defining these industries and providing the resources to help them grow. That includes having a flourishing ecosystem, with Chinese companies that will be highly competitive around the world. The government wants to see growth not only in manufacturing headsets but also in innovative technologies such as chips, screens, user experience,

3D modeling, motion capture, data processing, and positional tracking.

As one observer noted:

> "In 2018, China's Ministry of Industry and Information Technology issued an outline for the VR industry named 'Guiding Opinions on Accelerating the Development of the Virtual Reality Industry.' The document highlights the necessity of R&D and content service supply as a baseline for VR development. Its first developmental goal—constructing a robust VR supply chain—was reached in 2020 after building the Beidouwan VR Town in Guizhou province. Beidouwan VR Town is anticipated to reach a yearly production capacity of 1.5 million VR hardware pieces and total revenues of $145 million.
>
> "The second developmental goal is to see product supply, platform construction, and content production technology implemented in healthcare, education, manufacturing, and commerce. To help meet this goal, Microsoft launched the Nanchang City AI+VR Innovation Center, a cloud and mobile technology incubator, in Jiangxi province in 2019. Jiangxi's provincial government is optimistic of the potential to lure VR tech companies into this incubator in the hope of supporting and providing training for local companies. Both developmental goals are to be aided by professional training, the building

> of an industrial development base, and subsequent brand building once a dominant headset/hardware is identified.
>
> "Baidu...Alibaba...and Tencent...collectively known as BAT, are among the front-runners in China's VR/AR community.... In contrast with companies like Sony or HTC that have focused more on hardware development, BAT has acted primarily as a market facilitator. Because these companies lack the core technology specialties needed to build VR hardware capable of providing consumers with a three-dimensional experience, they instead focus on acquiring stakes in VR startups. By opening their platforms to content creators, the strategy is to wait for the dominant headset to emerge."[69]

However, Chinese VR and AR companies still depend on imported chips, and the country has yet to produce as much high-quality content as the U.S. But in China, entrepreneurs and the government express a confidence that with enough government support and market incentives, China will have the ability to catch-up and surpass other nations.

The Future of VR and AR

The U.S. is currently the global leader in VR algorithm technology and content in both the industrial and consumer environments. U.S. companies, including Facebook and Apple, will likely be active in developing VR hardware in the future, but most manufacturing will be located in Asia.

Apple was scheduled to introduce its new consumer AR glasses by 2023. The marketing power of Apple combined with the content support provided by the company was expected to result in its AR glasses becoming a high-growth area. Apple's strategy was to link AR glasses to iPhones, which would stimulate sales of both devices.

In addition to revenue from the AR glasses, Apple would realize significant sales from the AR content applications in its App Store. This could support the establishment of a large AR user base and a large community of software development partners. But a key challenge for Apple and other companies was to provide AR glasses with a lightweight and attractive appearance that could support a battery life of at least twelve hours. These kinds of technical requirements were difficult for Apple to meet and caused delays in the glasses' release.

South Korea's Samsung is expected to be a global leader in providing VR hardware, including ultra-high resolution displays and other key components. South Korea will emphasize exports. Israel, a leader in signal processing and compression technologies for VR headsets, is also likely to become a major creator of AI algorithms for VR.

China will place an increasing priority on establishing a VR ecosystem because of the large market for commercial and consumer applications and its desire to dominate cutting-edge technologies. China is setting up large and highly automated manufacturing supply chains for VR headsets and cubicles. This will allow the country to develop an increasingly large market share of the supply chain for VR devices.

China will also become a leader in the development of content and applications for the domestic market, starting with its current, impressive base in gaming content. (Tencent was the largest gaming content provider in the world in 2020.) But it is unclear at this time how successful China will be at creating content and services for VR markets abroad.

Today, a person enters the virtual world by wearing a headset and gripping hand controllers. In the future, there will be multiple ways to gain access to extended reality, such as cubicles with multiple displays and sensors on the walls and ceilings. As the user interface of virtual worlds becomes more sophisticated, users will also be able to use spoken commands, eye movements, and hand gestures to control their virtual reality experience.

There will also be a growing use of digital avatars, representations of the user that might look and sound like the real individual. An avatar might also represent a "perfected" person, eliminating characteristics they think might limit them in public. Or they can appear in the virtual world as a totally different person.

Despite the large range of games being developed to support new VR devices and the growth in device sales, the amount of VR hardware being used by consumers is still relatively small. But if VR does become a mass-market product, the industry will face a number of technical challenges.

For one thing, the amount of data that needs to be processed, stored, and made available to support VR is extremely large. Data centers that exist today do not have nearly the capacity necessary to support billions of people with advanced VR systems. Companies including Alphabet, Amazon, Microsoft, and Facebook as well as Alibaba, Tencent, Huawei, and Baidu are likely to establish cloud infrastructures for AI-based VR ecosystems. In addition, the real-time demands of advanced VR systems will require advanced high-bandwidth connectivity capacity. They will at least need 5G connectivity.

Then there is the issue of content. In the next decade, the sales volume of VR headsets is likely to reach fifty million to one hundred million units per year. That will be enough to broaden the content that is available for gaming and business applications as well as for entertainment. Access to additional content will, in turn, spark a virtuous cycle of more headset purchases leading to more content creation. The growth in headset sales will also stimulate enhancements in displays, fusion processors, and AI algorithms. The companies in this industry will potentially be very profitable.

My view of the future assumes that a number of countries, including the U.S. and China, will, in the long-run, have roughly equal capabilities in VR and AR, contributing to a relative equilibrium in the ways these technologies empower nations. But if one country takes a strong lead, it could have a destabilizing impact on the global balance of power.

A Dystopian Future?

The chances are that VR will have a positive effect on mankind's abilities at work and play. Combined with autonomous transportation in particular, VR and AR will be able to enhance people's performance and even their productivity on the road. Although it may make some jobs and industries obsolete, VR will create new ones. But it is also easy to imagine a future for VR that is dark indeed.

Suppose AI in its various forms, including robotics and VR, drastically reduce opportunities for employment. Technical improvements in VR will make it possible for many people to live in virtual worlds, almost to the exclusion of the real world. In their virtual worlds, people would lose track of day and night. Far more effectively than drugs or revolution, VR could satisfy the unemployed masses through distraction and mental pacification. In the worst-case scenario, the masses would be subject to government mind control.

In this virtual environment, many of the physical possessions that are part of daily life today will no longer be needed. If you need to own a painting, for example, you will purchase it as a virtual object, which can be sold or swapped, but never be touched. Eliminating the need for real goods could have a major impact on the structure of society. Many factories that produce physical goods will not be needed. Transportation requirements for goods and people will be reduced. As people withdraw into their own virtual realities, the "real" world might become a far emptier and hollower place.

Chapter 7

Smart Objects: 5G Wireless and the Internet of Things

In telecommunications, the fifth generation wireless standard, or 5G, will act much like a circulatory system delivering AI data to all parts of society. 5G is one of a number of technologies that are critical in making AI work in the real world. China has a lead in 5G as in many others areas.

Together, 5G and AI have the potential to transform a range of industries and achieve things like autonomous cars, long-distance surgery, and automated smart factories. In addition, AI and 5G together can help create the Internet of Things (IoT), a world in which the ordinary objects of our daily lives—from refrigerators to thermostats to automobiles—can transfer data to each other on wireless networks without human intervention. The IoT, gaming, and a variety of industries will be generating exponentially more data than ever before. It will take 5G's data-carrying capacity to handle it all.

5G can transmit data ten to one hundred times faster than 4G, the previous standard. That makes it possible for a phone to download a full-length movie in about fifteen seconds, compared with roughly six minutes on 4G. 5G can also drastically reduce latency—the lag in transmitting information from one point to another in a data network. As a result, a 5G-empowered device like a car, phone, or watch can communicate with the cloud or with other 5G devices almost instantaneously.

Moreover, the combination of 5G and AI will distribute intelligence widely, bringing AI closer to the end user and minimizing risks to data privacy and cyberattacks.

The nation that dominates in 5G will be able to lead in areas as diverse as logistics, digital health, smart farming, and quantum communications. And 5G will play an increasingly important role for the militaries of China and the United States. For example, if a tank or plane needs to exchange AI data with another tank or plane, it will require 5G to do it.

In 2021, China's lead in implementing 5G was considerable. At year's end, China had installed an estimated 1.2 million 5G base stations, the stations that transmit 5G radio waves to users. By contrast, less than 100,000 base stations had been installed in the U.S. by the end of that year.

China's early successes in the race for 5G implementation was the result of government policy and the capabilities of its political and economic system. In China, the government was able to set goals, give subsidies to carriers, and assign bandwidths relatively quickly. It established a timeline for 5G's implementation that began with scientific research and included creating new markets for 5G services.

By contrast, the story of 5G in the U.S. is one of private sector players making decisions based on shorter-term profitability forecasts and current market forces. AT&T and Verizon initially made some serious missteps. And during the Trump years, the U.S. government played a confused and sometimes conflicted role over whether to allocate adequate space on the spectrum to 5G carriers.

What 5G Can Do

The combination of AI and 5G will have a huge impact on the productivity of many industries. ABI Research estimates that the combination of AI and 5G will contribute about $18 trillion to the global GDP by 2035. Industrial manufacturing, transportation, and the retail and wholesale trades are among the key markets that will be impacted most. But the combination of AI and 5G will also help boost the productivity of other industries. Here are some examples:

Healthcare: AI and 5G have the potential to transform healthcare in a multitude of ways. The market for wearable tech that monitors everything from a heartbeat to blood sugar is booming, and caregivers are receiving previously unseen insights into the everyday health of their patients. With AI and 5G, healthcare providers will be able to combine the now available data with other data that impacts health, such as environmental factors like air quality. Today such information is not available for the average interactions between doctors and patients.

That holistic and real-time approach to healthcare is poised to transform everything from fine-tuning pharmaceutical doses in clinical trials to digitally delivering personalized patient care. 5G will also open the door to integrating new sources of data into personal care, like voice and video inputs.

Communities on the geographical fringes of healthcare will also feel the benefits as commercial 5G networks come online in rural areas that currently have limited access to healthcare resources. Today, patients with acute or complex conditions must travel long distances to get the care they need.

5G networks will also open up the frontier of "digital therapeutics," where therapies will be dispensed at home with the help of VR, AR, and mixed reality. Intel has been working on providing VR-assisted therapy for children with autism, and research has already shown that VR can help patients struggling with mental health and substance abuse issues.

Manufacturing: In manufacturing, 5G and AI will play a critical role in enabling highly automated factory floors, remote-controlled machines, recyclable infrastructure, mobile robots, and improved logistics. At some point, no humans will be needed to run a production line.

5G and AI will allow a wireless factory to reconfigure its production lines more easily and shorten production lead times. Currently, rearranging a production line to manufacture a different product requires hours or days of downtime as workers move and reconfigure machines. That process will be far faster, easier, and require less labor when there are no wired connections.

Media and Entertainment: Media content producers are currently researching and developing interactive content and video in which the user is presented with choices that directly affect the flow of programming. There is a clear opportunity for the combination of 5G and AI to collect behavioral consumer feedback to help content producers tailor shows to create more popular and captivating content. Interactive content will also mean there will be a need for higher bandwidth in the network, increasing the need for 5G capacity. And telecommunications providers will be able to better predict traffic spikes since video constitutes the lion's share of mobile traffic. Finally, 5G and AI will allow greatly improved encryption to prevent hacking and content tampering.

Retail: With buying things by voice becoming a growing trend as a result of products like Amazon Echo and Google Home, robots can offer a similar experience in a physical store. Cellular connections through 5G with the assistance of AI can improve the functionality and reliability of in-store robotic retail assistants controlled by either voice or touch.

Targeted advertising at the retail shelf is less intrusive and more pleasant for consumers, while being more effective for advertisers. Using AI and the increased penetration of 5G networks, targeted advertising on a shelf where the customer is standing can reach the customer just when she is making a purchase decision.

Another possibility being explored by retailers is preventing long lines at checkout stations through a combination of predictive analytics, machine learning, and traffic counting. AI algorithms can predict wait times by combining various data points such as historical traffic patterns, weather conditions, holidays, and special events.

Smartphone Volume

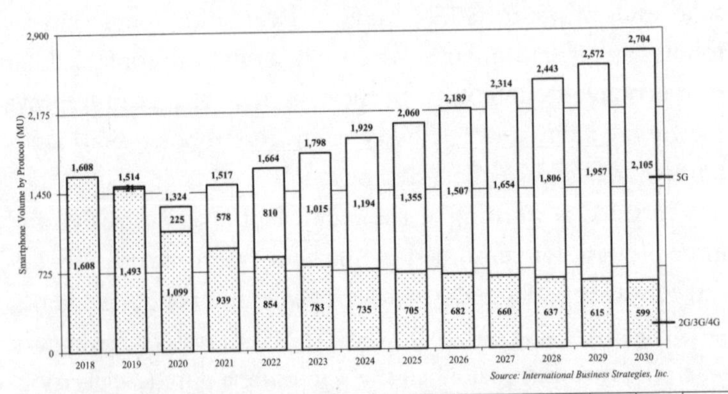

	2018	2019	2020	2021	2022	2023	2024	2025	2026	2027	2028	2029	2030
4G and below (MU)	**1,608**	**1,493**	**1,099**	**939**	**854**	**783**	**735**	**705**	**682**	**660**	**637**	**615**	**599**
Growth rate (%)	NA	(7.2)	(26.4)	(14.6)	(9.1)	(8.3)	(6.1)	(4.1)	(3.3)	(3.2)	(3.5)	(3.5)	(2.6)
Percent total (%)	100.0	98.6	83.0	61.9	51.3	43.5	38.1	34.2	31.2	28.5	26.1	23.9	22.2
5G (MU)	--	**21**	**225**	**578**	**810**	**1,015**	**1,194**	**1,355**	**1,507**	**1,654**	**1,806**	**1,957**	**2,105**
Growth rate (%)	--	NA	971.4	156.9	40.1	25.3	17.6	13.5	11.2	9.8	9.2	8.4	7.6
Percent total (%)	--	1.4	17.0	38.1	48.7	56.5	61.9	65.8	68.8	71.5	73.9	76.1	77.8
TOTAL (MU)	**1,608**	**1,514**	**1,324**	**1,517**	**1,664**	**1,798**	**1,929**	**2,060**	**2,189**	**2,314**	**2,443**	**2,572**	**2,704**
Growth rate (%)	NA	(5.85)	(12.55)	14.58	9.69	8.05	7.29	6.79	6.26	5.71	5.57	5.28	5.13

Source: International Business Strategies, Inc.

These graphics illustrate the rate of adoption of 5G smartphones.

Data Traffic

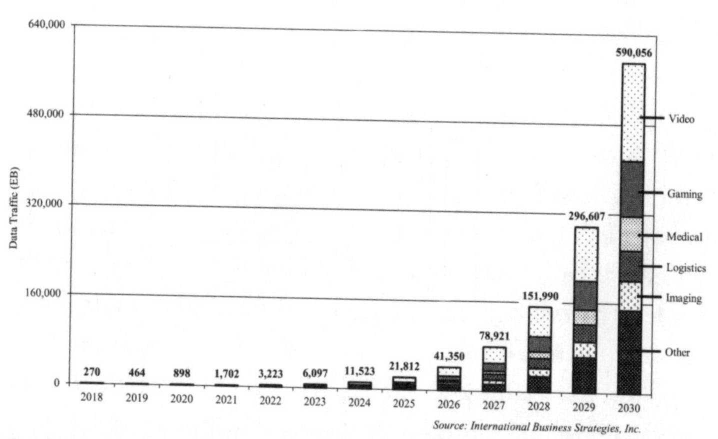

The figure shows the dramatic growth of data communicated between devices such as smartphones, personal computers, and data centers. Video and gaming are the largest contributors to data traffic, but new, high-growth generators of data, such as medical and logistics, are emerging.

5G in China

When 2G was introduced in the 1990s, China was nowhere to be seen. The key developers were Europe and the U.S., so the development and ratification of the standards were under their control.

When 3G was introduced by Europe and the U.S. in 2001, China decided to have its own version called TD-SCDMA. But China only had a limited ability to develop competitive 3G technology. Belatedly, China decided to adopt the 3G version of Europe and the U.S.

In 4G, China was part of the development of the specifications. But China lagged Europe and the U.S. in the installation of the 4G network by three years. Nonetheless, China put a great deal of effort into producing 4G smartphones, with Huawei becoming the world's second-largest 4G smartphone vendor. China also led the world in building the 4G infrastructure based on Huawei's technology domestically and in other countries.

With 5G, China was again part of the development of the specifications, and Chinese companies, specifically Huawei, led in 5G patents. China also decided to become the global leader in installing 5G infrastructure. It made rapid progress until the United States took actions against Huawei, which limited the supply of key components. These actions by the U.S. have slowed down but not stopped the company.

The Chinese government's commitment to 5G development included providing 5G licenses to China Mobile, China Unicom, China Telecom, and China Broadcasting Network to establish nationwide 5G network coverage. The government also provided a stimulus package for 5G-based services such as autonomous transportation, telemedicine, and ultra-high definition video.

Consequently, as more 5G base stations are installed, China's 5G infrastructure will operate at higher levels of utilization.

The Chinese government and Communist party realized that high-bandwidth and low-latency wireless connectivity would fire growth in industries such as digital health, autonomous transportation, logistics, factory automation, and interactive gaming. This commitment was similar in breadth to the long-term time horizons involved in China's successful plan to build over 25,000 kilometers of high-speed rail lines by 2020.

China is projected to install more than eight million 5G base stations by the end of 2025. Since 5G is considered strategically important for so many industries, the Chinese government and the state-owned enterprises are willing to make the investments necessary for China to become the world's leader in 5G—investments altogether totaling $100 billion or more.

The Chinese government is willing to invest in the building of 5G infrastructure even before there is demand for the technology. The government realizes that it will take a number of years before there is positive financial payback. But that is considered a necessary cost of business in the process of achieving global leadership in key technologies.

To ensure there is large demand for 5G bandwidth capacity in the 2025 to 2030 time frame, China is making investments to stimulate the growth in many emerging industries that will require high bandwidth, low latency, and support of AI. China is also running trials of applications to encourage different industries to use 5G. For example, Huawei has touted examples of 5G enabling remote diagnosis of COVID-19. In another case, Shandong Energy Group, a state-run mining company, announced the launch of a 5G network that would beam signals deep into underground coal mines.

The government's expectation is that its $100 billion investment will eventually pay dividends in the creation of new industries, new jobs, and opportunities as a result of entrepreneurs starting new companies. And the government hopes that building the 5G infrastructure capability in China will result in domestic 5G smartphone manufactures to gain market share, first in China and then in the global market.

China is also working on related communications technologies. For example, China has built an ultra-high speed quantum communications landline between Beijing and Shanghai. Quantum communications takes advantage of the laws of quantum physics to protect data. These laws allow particles—typically photons of light for transmitting data along optical cables—to take on a state of superposition, that is, they can represent multiple combinations of ones and zeros simultaneously. The particles are known as quantum bits or qubits. If a hacker tries to observe qubits in transit, their super-fragile quantum state collapses to either ones or zeros. This means a hacker can't tamper with the qubits without leaving behind a telltale sign of activity.

China has already begun work on the next generation of wireless technology, 6G, which will have ten times more bandwidth than 5G. China plans to develop 6G without being dependent upon American technologies. By making investments in research at places like the Chinese Academy of Sciences and engaging private corporations, including Huawei and ZTE, China hopes to build on its lead in 5G technology to support the rapid adoption of 6G technology by 2030. With China ahead of the U.S. in 6G, the technological gap between the two countries will only grow, with many more AI-based applications available in China than in the U.S.

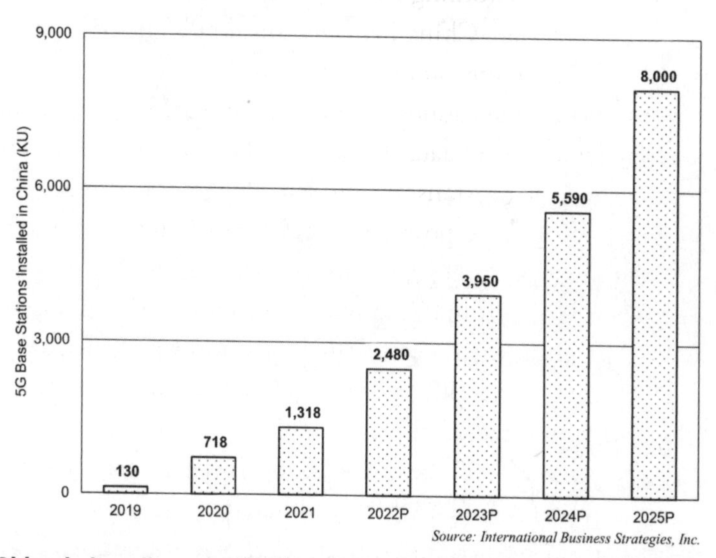

China is heavily committed to global leadership in the adoption of 5G, which required an investment of $100 billion in establishing the 5G infrastructure. The Chinese leadership plans to install eight million base stations by 2025.

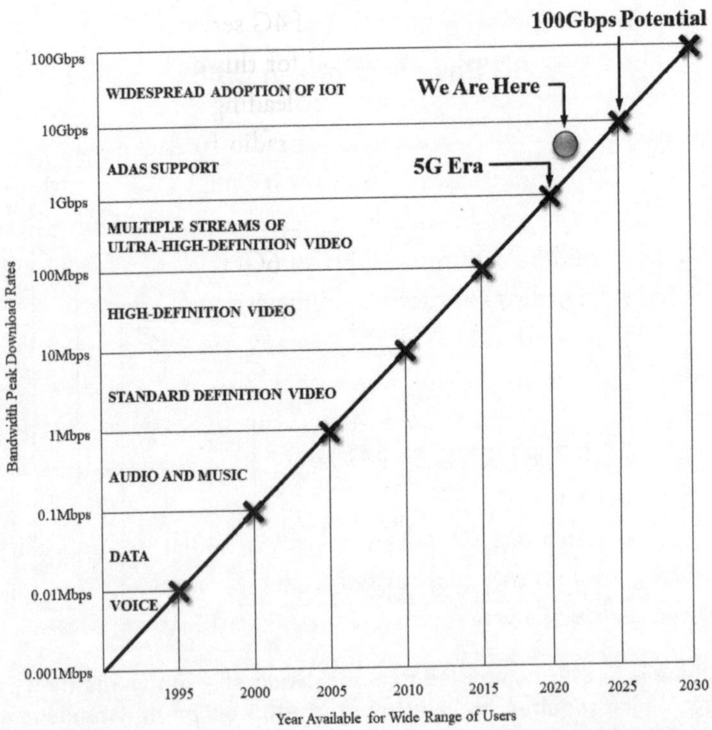

China is active in developing 6G, which will expand the bandwidth of wireless connectivity dramatically. It is expected to install initial 6G capabilities between 2028 and 2030. China's position in 6G intellectual property will lead that of the United States, which has made far less of an investment in 6G technology.

5G in the United States

For most Americans in 2021, the reality of 5G had not lived up to the hype. U.S. wireless carriers advertised as early as 2019 that they had nationwide 5G in place. But two years later, it was nearly indistinguishable from that of 4G service.

There were a variety of reasons for this delay: But a major factor was poor decision-making by leading American carriers. One decision involved the choice of radio frequencies to carry 5G traffic. The radio wave spectrum includes a range of radio frequencies that can be used by 5G carriers: a low-band (below 1GHz), a mid-band (from 1 GHz to 6GHz), and a high-band (over 24GHz), also known as "millimeter wave."

China, South Korea, and Germany are focused on mid-band. It is the sweet spot that strikes a balance between fast transmission and long signal ranges. But in the U.S., the two largest carriers, AT&T and Verizon, at first focused on using the millimeter wave.

That was a major strategic mistake. While 5G millimeter waves allow for faster transmission of data than mid-band or low band, they have a very short service range—under one mile. And millimeter waves are prone to interference from rain and snow. Millimeter waves also do not penetrate walls. As a result, the carriers were faced with the task of building a large number of short-range cell towers in dense cities. Even there, however, it would be necessary to construct complementary Wi-Fi infrastructures to carry the millimeter transmissions inside of buildings.

By contrast, there are many advantages to mid-range signals. They support a service range of several miles. A person using a mid-range 5G smartphone could continue to have connectivity while walking down the street, sitting inside a building, or riding in a car or train. Mid-range signals transmit data more slowly than millimeter signals, but they can provide a smartphone user with more interactive gaming than 4G. Mid-range signals can also provide ultra-high definition content for use by televisions, tablets, notebooks, desktop computers, and smartphones.

Another reason for the delay in the implementation of 5G in the U.S. was that the federal government resisted freeing-up mid- and low-band 5G. Those parts of the spectrum were owned or coveted by satellite companies, cable companies, and broadcasters. During the Trump administration, a number of industries lobbied intensely in Congress and at federal agencies.

In addition, the 5G implementation process was slowed by the business priorities of the major carriers. AT&T deprioritized establishing the 5G infrastructure because of its focus on the development of its entertainment business. Without competitive pressure from AT&T, Verizon felt little impetus to accelerate its 5G network's expensive build-out. For the most part, it could simply optimize profits from existing 4G technology. Without significant pressure from the government or a competitor, AT&T and Verizon simply did not feel there was any rush.

A number of factors changed these dynamics in 2021. After the election of President Joe Biden, the federal government began to auction off space on the radio spectrum for use by 5G carriers. Meanwhile, as a result of the merger of Sprint and T-Mobile, T-Mobile emerged as a major challenger offering a growing network using mid-range wavelength 5G. New leadership at Verizon began to commit to major investments in mid-range 5G technology. And AT&T, now freed of its entertainment division, decided to make a major commitment to mid-range 5G as well.

In January 2021, the government held an auction of the mid-range radio wave spectrum, in which wireless carriers bid a total of $81 billion. Verizon bid $45 billion, the highest at the auction, for the spectrum licenses. AT&T came in second, bidding $23 billion. And T-Mobile had the third highest bid of $9 billion. The competition was on, and the U.S. was now destined to receive mid-range 5G service.

The Internet of Things (IoT)

The Internet of Things refers to an enormous network of wirelessly connected machines. The IoT will encompass billions of network-connected sensors, cameras, wearable devices like AR glasses, vehicles, assembly lines, and just about anything else, from dog collars to robots. Combined with the computing power of the cloud, IoT delivers insights and new levels of machine-to-machine coordination that could improve the quality of life and accelerate economic growth. According to Business Insider, there will be more than sixty-four billion IoT devices installed around the world by 2026.

By 2030, 80 percent of generated data will be processed at the edge, compared with 20 percent in 2015, and the market for IoT devices will be $252 billion.

The IoT world will be inhabited by devices we already know, such as Amazon Echo, Google Home, Apple Watch, and Fitbit. The use of 5G in car-to-car communications is still in its infancy. But it also has the potential to make driving safer, more convenient, and eventually autonomous. Sensors in cars will be able to do everything from a remote start to popping open the trunk. Other applications will include:

Smart Homes: The IoT will allow homes to be smart and secure with sensors, alarms, cameras, lights, and microphones communicating with one another and providing security twenty-four hours a day. Such security systems allow users to see, hear, and speak to pets at home and visitors at the front door by their computer, tablet, or mobile phone.

Self-Healing Machines: Manufacturing equipment can be designed to recognize variances in their own operations and correct them, using arrays of thousands of sensors as well as AI and machine learning. This allows companies to solve problems before they turn into issues that require repair and downtime. This, in turn, saves companies money and frees up employees who would normally monitor equipment and undertake maintenance to work on higher-level tasks.

Ingestible Sensors: An electronic device the size of a pill is outfitted with a battery, a microprocessor, and sensors. It can be swallowed to monitor conditions inside the gastrointestinal tract to detect, among other things, bleeding and absorption of pharmaceuticals.

Motion Detection: Detecting vibrations or motion with sensors in large-scale structures like bridges and dams can identify small disturbances that could lead to catastrophic failures. Networks of detectors can also be used at locations likely to experience earthquakes, avalanches, and landslides.[70]

Smart Contact Lenses: Contact lenses that can collect health information or treat specific eye conditions are currently the subject of considerable research. The Swiss company Sensimed, for example, has developed a noninvasive smart contact lens called "Triggerfish" that has a sensor embedded in a soft silicone contact lens. It can detect tiny fluctuations in an eye's volume, which can be an indicator of glaucoma. The device transmits data wirelessly from the senor to an adhesive antenna worn near the eye.

Smart Farming: An increasing number of farmers will be using IoT-enabled tools to monitor soil conditions, soil moisture levels, crop health, and livestock activity. The data can be analyzed to determine the best time to harvest plants and create fertilizer composition and schedules. Farmers can also take advantage of drones to collect atmospheric data and photos.

Smart Cities: The IoT could transform cities by solving the problems citizens face every day. With sensors, connections to the cloud, and the right data, the IoT can reduce traffic jams, noise, pollution, and crime.

China is accelerating the adoption of IoT concepts for many applications and is already a global leader in areas such as smart cities, databases with 3D facial images, and health monitoring. Additionally, China and South Korea are leaders in establishing 5G infrastructures, which may give high-bandwidth support for data-centric AI activities. China is also making large investments to enhance the capabilities of AI algorithms that are needed for many high-growth applications, such as logistics, robotics, digital health, and autonomous transportation.

China's 1.4 billion people will generate enormous quantities of data compared to other countries. Currently, China's data center and cloud investments lag those of U.S. companies. However, this situation will change, and large investments will be made to expand the cloud and data center capabilities in China.

China is moving into the IoT arena at a faster rate than the United States and Europe. Large subsidies are being provided by the Chinese government to accelerate the development and adoption of these new technologies.

A key area of China's competitive advantage is the development of new markets to adopt new IoT concepts as well as developing the necessary technologies to generate data through IoT devices. Consequently, there will be a very large market for IoT devices in China, but with the need to acquire sensors and semiconductor functionality from outside of China.

Lessons Learned

The lessons from the first years of the competition for 5G are fairly easy to grasp: China, with its longer-term goals and ability to turn those goals into reality, had a distinct advantage over a system driven solely by market forces. What was lacking in the United States was government leadership in creating a profoundly important infrastructure that would not immediately generate profits for investors. Like the Eisenhower administration's decision to build an interstate highway system in the 1950s, a government-led program for the creation of 5G and 6G wireless networks would pay immense benefits to the economy and society in the long term.

Instead of establishing and implementing a long-term strategy to establish advanced technologies, the U.S. government is attempting to slow down China's progress by blocking some Chinese companies from gaining access to U.S.-made semiconductor products. This will have a negative impact on China in 5G and other areas in the short term. In the longer term, however, China will strengthen its supply chain through making large investments in developing the products needed to support 5G. When China builds a sufficient supply chain, Chinese companies will no longer rely on U.S. semiconductor companies. This process will certainly lock U.S. companies out of the Chinese market.

While Huawei's smartphone business has been seriously damaged by the U.S., other Chinese smartphone vendors have filled the void. The impact on the total supply chain of 5G smartphones in China is minimal. The result is that the total volume of smartphones provided by Chinese companies continues to increase and the overall benefits to the U.S. from the restrictions on China's smartphone production are minimal. However, the moves against Huawei have had an impact, building an increasingly negative attitude in China against the U.S. This change in attitude is stimulating the Chinese to accelerate their approaches to the development of new technologies.

China has Huawei, ZTE, and Datang Telecom Group as companies that can develop 6G technology. The United States, however, has to rely on Ericsson and Nokia, which are behind China in 5G technology by two years or more and will be behind China in 6G technology. By 2030, China will likely be ahead of the United States in wireless technology by five years or more.

Chapter 8

China vs. the United States: A Sprint or a Marathon?

In October 1957, a beachball-sized aluminum alloy sphere sailed across the sky in an elliptical orbit one hundred miles above the surface of Earth. As it flew overhead, it broadcasted a squeaking, wobbling beep that could be heard on the earth below by tuning in and listening to its low-frequency radio transmission.

When Sputnik, the world's first human-made satellite to orbit the earth, was launched into space by the Russian government from the Baikonur Cosmodrome in Kazakhstan, it came as a complete shock to most Americans. The launch came sixteen years after the attack on Pearl Harbor—a national tragedy that was still fresh in most Americans' minds. Pearl Harbor had created a lasting fear of a surprise attack by a hostile foreign power.

More ominously, the Sputnik launch had come just eight years after the first Soviet atomic weapons test, less than four years after the Soviets had detonated its first hydrogen bomb, and less than two years after the detonation of their first megaton hydrogen bomb.[71]

A little over a month after the launch of the first Sputnik satellite, in the predawn hours of November 8, 1957, a yellow-white light streaked across the sky above New York City.[72] The source of the eerie glow was the reflected sunlight, or satellite flare, from the Soviet's next iteration of Sputnik—Sputnik 2. What most Americans who remember Sputnik 2 recall, if they remember anything at all, was that Sputnik 2 carried the first space passenger within its capsule—a thirteen-pound stray dog named Laika. However, what was worrisome to the U.S. political and military establishment was the massive scale of this new satellite: Sputnik 2 weighed more than half a ton—five times the weight of the original Sputnik 1.

The sheer scale of Sputnik 2 came as a terrifying revelation to American scientists, engineers, politicians, and policy analysts. The ICBM (intercontinental ballistic missile) that had carried the enormous Sputnik 2 payload could be redeployed by the Soviet military to deliver a massive atomic or thermonuclear weapon to an American city. With the twin launches of Sputnik 1 and Sputnik 2, the space age—and the great space race between the two nuclear-armed global superpowers—began in earnest.

There are some distinct similarities between the space race with the Soviet Union and the development of artificial intelligence today in the United States and China. Both are high-stakes technology competitions. And both have enormous military, political, and economic implications.

During the early stage of the space race, the United States fell significantly behind the Soviet Union. Just twenty-three days before the American astronaut Alan Shepard's historic space flight, Russian cosmonaut Yuri Gagarin successfully completed one orbit of the earth, simultaneously beating the Americans to two records at once—becoming the first human in space a little over three weeks ahead of Alan Shepard and beating John Glenn in the race to become the first human to orbit the planet by nearly a year. The United States also lagged behind the Soviet space program in other key firsts, including being the first nation to launch a space mission with multiple crew members, the first country to successfully complete a spacewalk, and the first to put a woman in space.

We now know, of course, that the United States triumphed in the space race. The United States government took a leadership role and rallied the political, military, economic, and technological resources of the nation. As a result, within four years of Sputnik's first orbital flight, the United States launched its first Saturn rocket, the Saturn I, inaugurating the Apollo space program. As every schoolchild in the U.S. has learned, American astronauts first set foot on the moon on July 20, 1969, after their Apollo 11 spacecraft landed on the lunar surface at the Sea of Tranquility. The Soviet Union, by contrast, never landed a cosmonaut on the surface of the moon.[73]

Two questions now hang over the evolving marathon competition between the United States and China in the field of artificial intelligence: When, if at all, will America experience another Sputnik moment? Will the United States government be able to plan and execute AI development strategically?

3500 Years of Memory

When comparing the political and economic perspectives of the United States and China, one important thing to keep in mind is time frames. Thoughtful American political and business leaders might put today's current events in the context of the last century or in the time since America became an industrialized nation after the Civil War. Chinese leaders, by contrast, think in terms of more than three and a half millennia, since the establishment of the Shang Dynasty. During most of that time, China was the biggest, most populous, richest, and technologically advanced nation on Earth. Perhaps the best indicator of the Chinese perspective was the name it called itself: the Middle Kingdom. By this, the Chinese meant it was the most important and powerful nation in the world, located in the center of the universe. China invented such things as gunpowder, paper money, printing, a postal system, and the magnetic compass, far surpassing the technological innovations of competitors. Even as empires such as Rome and Byzantium rose and fell, China remained technologically ahead and, in Chinese minds, culturally superior. The emperors of the most powerful dynasties in Chinese history, the Yuan and the Qing, had been foreign invaders from Mongolia and Manchuria, respectively. But so powerful was Chinese culture that these invaders were sinified, adopting the Chinese language and the existing system of bureaucracy. In the end, China defeated these invaders by making them more or less Chinese.

In 1757, at the height of the Qing Empire, the emperor Qianlong imposed the Canton System on traders from Europe. All formal trade was required to pass through thirteen trading companies located in the city of Guangzhou, which the British at the time called Canton. Chinese citizens were forbidden from teaching the Europeans. In addition, European traders were not allowed to bring women to China. In retrospect, this move may be seen as a sign of growing insecurity on the part of China's leaders. For the first time, foreigners possessed superior military technologies and potentially threatened Chinese government hegemony. While China built a navy to trade with other countries, the emperors were very concerned with other cultures intellectually polluting China and were ignorant of the military technology progress in other countries.

During the following decades, the Chinese refused to sell their much-desired tea, silk, and porcelain to foreigners for anything but silver. As a result, the European's supply of silver fell and the price of the metal skyrocketed, and trade deficits rose. European nations, particularly the British, desperately looked for goods the Chinese would buy. They found their answer in opium, a profoundly addicting product that was produced cheaply in India. By 1833, the British were exporting 30,000 chests of 170 pounds each to China and the number of Chinese drug addicts totaled between four and twelve million.[74] The Chinese emperor Daoguang begged Queen Victoria to halt the opium trade, but his letters garnered no response. When the imperial court ordered the confiscation of all supplies of opium and a blockade of ships, the Royal Navy responded by sending a flotilla up the Pearl River to capture Canton and Nanking. Superior British military technology overwhelmed Chinese forces and the war ended in 1842 with China losing Hong Kong to Great Britain. Under duress, it also established five treaty ports for the British along the Chinese coast. The following year, the British secured the rights to extraterritoriality—its citizens in China were not subject to Chinese law, only to British law. The French government soon gained the same extraterritorial concessions from China. Over the following years, the Chinese government, powerless in the face of European military might, granted the establishment of eighty treaty ports open to the opium trade and the right of all Europeans to travel anywhere in China with extraterritorial freedom.

The time period stretching from the end of the First Opium War in 1842 through the victory of the Chinese Communists in 1949 is referred to by the current regime as the period of "the Great Humiliation." It was that. But it is important to remember that it was also a hiatus, a parenthetical, and a short episode in the long run of Chinese history.

In 1992, when Francis Fukuyama published his book, *The End of History and the Last Man*, he declared the victory of the western system of democracy and market capitalism as the default form of national government. To Fukuyama, who received much critical praise and attention at the time, democratic capitalism was the post-ideological and final future for nations. But his calculation was based on an observation of nations over a relatively short period of time—basically since the French Revolution of 1789. Chinese people understandably consider this analysis superficial and inaccurate. For China, history began thousands of years ago and most of it has consisted of a successful China—ahead of all other nations in wealth and power—ruled not by democracy but by an autocratic form of government.

Comparison of Per Capita GDP for China and the United States 1980 to Present

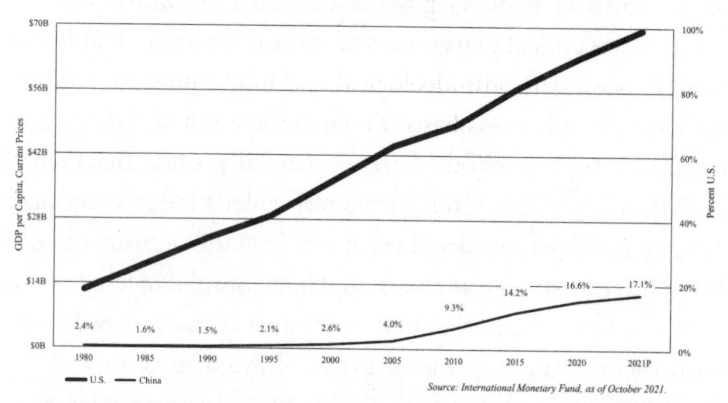

Source: International Monetary Fund, as of October 2021.

China's GDP will likely pass that of the United States by 2030, but its GDP per capita will be much lower. Nonetheless, China's GDP per capita is understated because much of its economy is cash and barter. In addition, the buying power of people in large urban areas of China is closing gap with the United States. The number of billionaires in China is growing rapidly.

China's Progress

The economic rise of the People's Republic of China since the death of Mao Zedong in 1976 has been nothing short of extraordinary. During less than fifty years of economic reform, the Chinese have lifted more than 800 million people out of poverty to become an upper-middle-income country. China's gross domestic product, one of the broadest measures of economic output, is now the second largest in the world, just behind that of the United States. Based on realistic calculations, the GDP of China will be higher than the United States by 2030, although GDP per person in China will continue to be much lower than in the United States. The reforms introduced by Deng Xiaoping in the late 1970s, known as "socialism with Chinese characteristics," looked little like anything in the history of Marxist-Leninism. What it did resemble greatly was the economic strategy undertaken by the five "Asian Tigers" of Japan, South Korea, Taiwan, Singapore, and Hong Kong in the decades following the Second World War. Those economies had drastically outperformed China's during Mao's rule during 1949 to 1976, as they transformed their nations from states of poverty and deprivation to world-class exporters of goods and services. The formula of the Asian Tigers seems remarkably familiar to observers of China since the late 1970s: strong, government-managed industrial policies that encouraged foreign investment and supported and protected nascent domestic manufacturing industries until they could compete on the world stage. The story is not so much a success story of free-market capitalism as it was a tale of successful government-managed industrialization.

The statistics are staggering:

- In 1979, China consumed 276.2 billion kilowatt hours of electric power; in 2020, it consumed 7.51 trillion kilowatt hours. Demand is now so strong that there are severe shortages of electricity in China.
- In 1988, there were 3000 mobile phone users in China; in 2020 there were 1.59 billion.
- Between 1991 and 2020, China's annual research and development spending grew 169 times, from 14.3 billion yuan ($2.21 billion) to 2.44 trillion yuan ($382.69 billion). China's total R&D expenditures overtook Japan's in 2013, making them the world's second largest after the U.S.
- China's applications for patents grew from zero filings in 1984 to 68,720 in 2020, overtaking the U.S. in 2019. China also surpassed the U.S. in terms of the number of academic research papers published in 2016.
- In mid-2021, China had installed 916,000 5G base stations, accounting for 70 percent of the world's total and serving more than 365 million 5G-connected devices. Work on 6G technology is underway, scheduled to be introduced in the late 2020s.
- More than 23,000 miles of high-speed rail lines have been built in China since 2008, including the Beijing-Shanghai rail line which speeds trains between the two cities at up to 220 miles-per-hour, reducing what used to be a 17-hour trip to 4.5 hours. A 370 mile-per-hour maglev train debuted in China in 2021.[75]

Government Spending on R&D: U.S. vs. China

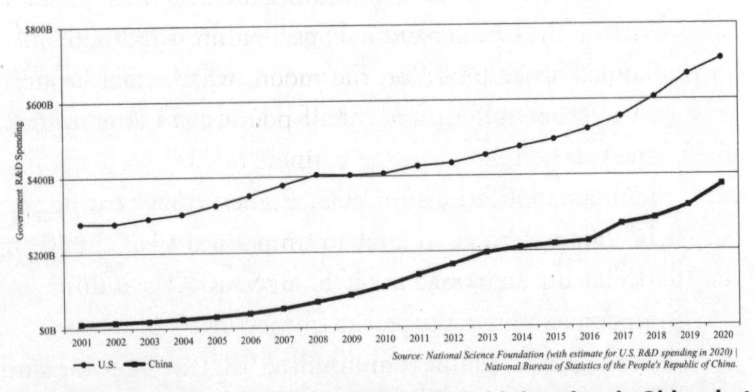

Government spending on R&D in the U.S. is higher than in China, but with lower wages in China, the number of people working in R&D is close to that of the U.S. Though the U.S. reports spending over $700 billion in R&D, new inventions seem to be limited, and most of these new innovations are created by the private sector.

Chinese cutting-edge technologies have spanned the sciences and a host of industries. They now include innovations in fields as diverse as financial technology, biotechnology, surveillance technologies, power grid cybersecurity, and AI-driven autonomous weapons systems. China is also pressing ahead in robotics while developing the market for home care robots, a market which could create the demand for hundreds of millions of new robots units in the future.

A glimpse of Chinese scientific progress could be seen in the autumn of 2020, when the Chinese announced two significant achievements. The China National Space Administration landed an unmanned space probe on the moon, which then brought back to Earth nearly four and a half pounds of volcanic rock. The lunar rock samples from the Chinese mission were the first from the moon's surface since 1976. Earlier[76] that year, China became the first country to land an unmanned space probe on the "dark side of the moon" at the lunar South Pole. China has also brought back samples of rocks from Mars.

Chinese researchers also announced a giant step forward in quantum computing. They had demonstrated "quantum advantage," a term used by scientists to describe using photons of light to solve problems that classical computers cannot solve, according to an article published in the scientific journal *Science* by the Chinese researchers. According to experts in the field, the Chinese researchers' work outpaced some aspects of Google's efforts to move quantum technology ahead of classical computing.

Production of Industrial Robots in China

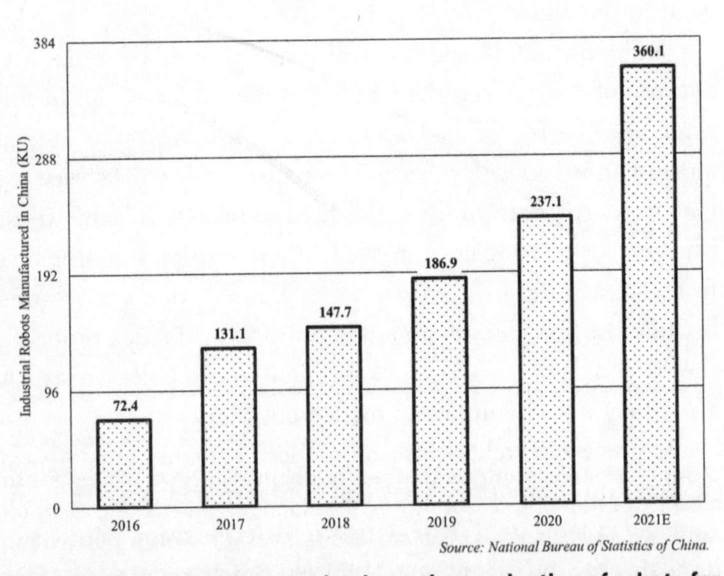

Source: National Bureau of Statistics of China.

China is placing a strong emphasis on the production of robots for industrial use and use in the service sector. Both the number of robots and the intelligence of robots in China is increasing. Robots will be used for home care of the aged, precision surgery, and logistics as well as in factories.

Jump Ahead Strategy

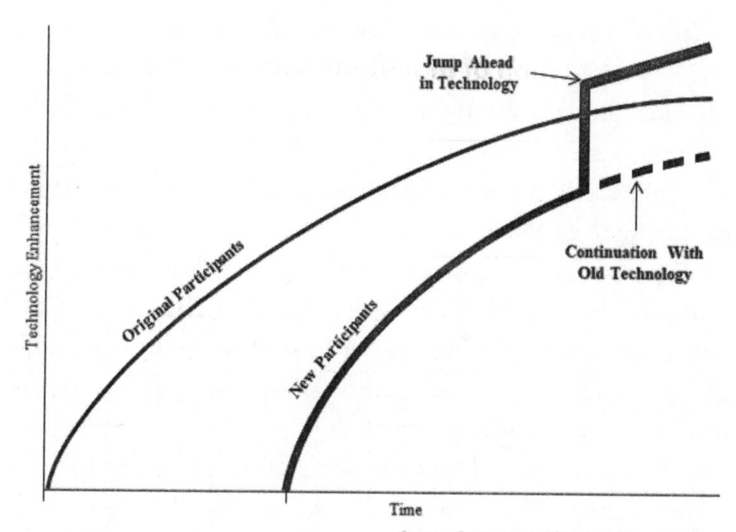

Source: International Business Strategies, Inc.

Companies and countries that lag in technology have difficulties in closing the gap with competitors if they focus on the existing technologies. To overcome this problem, China invests in future-generation technologies. The U.S. government used this *jump ahead* strategy in the past but abandoned it. Israel uses the jump ahead strategy in many areas with great success.

The Chinese government and industrial companies are collaborating on the jump ahead strategy in many areas. For example, a 600kmph train is scheduled to be in service in 2024 and a 1000kmph train is scheduled for 2026. China is pursuing a jump ahead strategy in AI both in military and commercial applications.

Competing Through Education

The Chinese education system is now structured to develop a large cohort of engineers who are highly skilled in AI. With over seven million new students graduating from institutes of higher education every year, China has already begun to lay the groundwork for developing an expert workforce. Chinese university graduates are also incentivized to perform academically by what is, in effect, a two-tiered system of professional rewards: In 2021, the top 15 percent of students in engineering and science earn more than double than that of other university graduates.[77]

As AI, engineering, and high technology have moved to the center of China's public policy goals, the selection process for the best talent has shifted ever earlier in students' education, with the sorting of top-tier talent now beginning long before students reach the university level.

The exams that determine which students go to the top Chinese universities are based on three specifically targeted academic subjects: mathematics, which demonstrates the ability to complete complex problems; physics, which demands mastery of complicated technical concepts; and Mandarin, which places powerful demands on students' memories. These subjects are also neatly aligned with the critical skill required of future engineers, business leaders, and bureaucrats in technology.

In order to develop the country's most promising talent, China has a rigorous selection process for students. The process begins at fifteen or sixteen years of age and is used to identify and select the individuals who are seen as having the greatest capacity for future achievement. These high-achieving students are then given the most intense level of education, especially for new technologies and business, while the lower achievers are given less-intensive training. In fact, even earlier in their education, before some American students have entered primary school, their Chinese counterparts have already started down the path of tech education. Chinese parents encourage their children to learn computer programming at five or six years old by attending after-school classes. Programming languages, such as Python, are considered even more critical for Chinese children to learn than English language fluency. Increasingly, schools are including physical fitness in the curriculum.

The Military

China experienced explosive growth not just in the technology sector but also in its rapid military development. The People's Liberation Army now boasts the largest standing ground force in the world. These armed forces possess the largest navy globally when measured by its total battle force of ships and submarines. China has over 1,000 ballistic missiles with ranges above 500 kilometers, while the longest-range Chinese missiles boast a maximum range of over 5,000 kilometers. China already possesses ground- and sea-based nuclear weapons systems; moreover, the People's Liberation Army is currently working toward improving those existing weapons systems and pursuing nuclear-capable, air-launched ballistic missiles to create a full nuclear triad for strategic deterrence. The Chinese military is also keenly aware of the importance of cyberwarfare capabilities, and they have deployed a portion of the world's second-largest defense budget to develop their cyber capacity accordingly. In deference to the emerging centrality of cyberwarfare, a commonly echoed adage among Chinese military analysts is that cyber warfare is strategic warfare in the information age, as nuclear warfare was in the twentieth century.

Reconstructing the Middle Kingdom: the Belt and Road Initiative

The Belt and Road Initiative (BRI), which the Chinese government began in 2013, is a colossal international undertaking. The initiative, which has an estimated cost of over a trillion dollars through 2027 and is scheduled to run through 2049, is analogous in some ways to the Marshall Plan that the United States established after World War II. The BRI attempts to expand China's trade reach and economic power by tying together swaths of Asia, Europe, and Africa—uniting them (in some cases to become Chinese client states) within the BRI network. The initiative is establishing a massive cross-border transportation infrastructure to allow goods to be moved more efficiently over a vast trading area. The initiative will provide outlets for Chinese-made goods and services and the rapid delivery of oil, gas, and other commodities from their production hubs to China.

The BRI has made major inroads in Europe. Between 2008 and 2019, Chinese companies made nearly 680 deals, valued at about $255 billion in thirteen European countries. Among the 160 companies completely taken over by the Chinese were Pirelli, the Italian tire maker, and thirteen professional soccer teams. Chinese state-owned companies focused on buying stakes in European ports, including Antwerp in Belgium; Le Havre, Nantes, and Marseilles in France; Genoa in Italy; Bilbao and Valencia in Spain; Rotterdam in the Netherlands; and Piraeus in Greece. China has also established a railroad system for transporting goods between Xi'an, Chengdu, and other cities in China and Europe.

The BRI has run into some headwinds. With many of China's corporations and other priorities weighing on the Chinese government, China may be struggling to fully fund the BRI. In addition, some nations, including a number of Eastern European nations, have expressed growing resentment for what they say is a heavy-handed and domineering approach to BRI-funding recipients.

But Chinese leaders say the long-term goals of BRI will eventually be achieved using the trillions of dollars projected to be spent by 2040. It is a small price to pay to make China once again the Middle Kingdom, the most important and powerful nation in the world, located in the center of the universe.

A Call to Action

The events of the past few years are not encouraging. The Chinese Communist Party and the government of China since 2015 have made remarkable progress in the development and implementation of an array of new technologies, including artificial intelligence. In 2015, the Communist party announced the Made in China 2025 initiative. It outlined a series of national industrial policy goals that would take China from a labor-intensive manufacturing economy to leadership in critical areas of high technology within a decade. The goals it outlined included rapid progress in AI, machine learning, electric vehicles, green energy, aerospace, high-tech ships, railway equipment, agricultural technology, medicine, and medical devices.

The United States must act now to plan, manage, stimulate, and support development of AI and related technologies within its borders. There is no time to lose. By 2030, China will be firmly ahead of the United States in AI. In 2040, major parts of society will be under the control of AI, and it will be too late.

The United States is superb when it decides to take on the responsibilities of global leadership. For precedents, one can look at Franklin D. Roosevelt's mobilization for World War II and Dwight D. Eisenhower's initiatives to buttress the national defense in the 1950s with the assistance of the best administrators and scientific minds available in the nation. And, of course, one can look at John F. Kennedy's commitment to the space race. These initiatives not only rapidly achieved their goals, but they also spawned a host of new technologies that made possible, among others, the microprocessor, the internet, smartphones, and GPS.

I am not a politician, political scientist, or expert in government. I am just a man who has worked in industries producing cutting-edge technologies for many years. So I think it would be unwise for me to try to describe in detail how the government of the United States can best organize a national AI initiative. But I have some ideas of where to start:

- **Have the government lead:** Establish a team at the top levels of the federal government consisting of experts who have the technical, business, and social expertise to understand the opportunities and threats related to AI. Currently, national security is defined in terms of diplomacy and defense, not economics or technology. For example, the National Security Council deals with the Defense and State Departments. It has little contact with the National Economic Council and the Departments of Treasury, Labor, Agriculture, and Health and Human

Services, much less the Federal Reserve and the Security and Exchange Commission. The administration should consider establishing an AI security council, drawing off the expertise of all areas of government. What we need, in the words of the Brookings Institution, is a whole of government strategy.

- **Focus government attention on deficiencies in the American defense and the aerospace industries:** Establish new laboratories and strengthen existing ones to develop new AI-based algorithms for the defense and aerospace sectors. We need to assure military superiority in all our armed forces and assure they use the latest generation of semiconductors and components, not those that are twenty years old. New software will need to be developed based on the latest generation of AI-related technologies.
- **Focus government attention and resources on key industries:** The federal government's lead entity on AI needs to set a small number of priority areas upon which to focus its attention. Priorities should be selected after a careful analysis of AI development in China and other countries. Potential areas for prioritization should include:
    - digital health
    - autonomous transportation
    - logistics
    - the digital food supply
    - the future role of digital twins
- **Invest in research and development:** The Chinese government invested $376 billion in R&D in 2020, up 10 percent from the year before. It plans to keep increasing

R&D investments by 7 percent annually through 2025. The federal government invested less than $60 billion in 2020. We have got to up our game. One strategy to consider is to increase federal R&D tax credits to stimulate corporate research.

- **Fund and support key AI-related initiatives in the private sector:** Chinese technology is being developed today with funding from both the private sector and the Chinese government. In the United States, funding depends upon the private sector alone. This puts the United States at a great disadvantage, particularly in areas where the new technology will not immediately generate profits. Emphasis should be placed on making the United States a future exporter of new technologies.
- **Focus business and the capital markets on long-term investment and planning:** Since the 1980s, companies have increasingly focused on return on capital at the expense of long-term R&D and technology development. Most corporate shares are held for less than a year. And corporate leadership and capital markets tend to focus on quarterly stock performance rather than the long-term goals. Ideas to consider are implementing tax incentives for holding equity positions longer and developing new metrics to measure long-term corporate performance that includes the value of R&D.

With its form of governance, China's leaders have been able to map out long-term plans for development of technologies by coordinating the private sector, state-owned industries, think tanks, universities, and government bodies from the national to the local level. Meanwhile, the United States has depended on its private sector and market forces to lead the way in the development of artificial intelligence. With the two national systems pitted head-to-head, America is falling behind. This is not for want of great American companies, thinkers, creators, or innovators. It is the result of a lack of leadership. What America needs now is the kind of leadership that brought success in the space race—the kind of strategic leadership that only the federal government can provide.

We need another Sputnik moment now because the United States—if it loses the lead in artificial intelligence and related technologies—will experience a devastating blow to its economic competitiveness, military preparedness, political influence, and to its citizen's quality of life.

Endnotes

1 Graham Allison and Eric Schmidt. "Is China Beating the U.S. to AI Supremacy?" Belfer Center for Science and International Affairs, August 2020, www.belfercenter.org/publication/china-beating-us-ai-supremacy.

2 Cade Metz, The Genius Makers: The Mavericks Who Brought A.I. to Google, Facebook, and the World (New York: Random House, 2021, pp. 226-227).

3 Prashant Loyalka, Ou Lydia Liu, Guirong Li, Igor Chirikov, Elena Kardanova, Lin Gu, Guangming Ling, Ningning Yu, Fei Guo, Liping Ma, Shangfeng Hu, Angela Sun Johnson, Ashutosh Bhuradia, Saurabh Khanna, Isak Froumin, Jinghuan Shi, Pradeep Kumar Choudhury, Tara Beteille, Francisco Marmolejo, and Namrata Tognatta, "Computer science skills across China, India, Russia, and the United States." PNAS, National Academy of Sciences, April 2, 2019, www.pnas.org/content/116/14/6732.

4 "Wright Flyer," Wikipedia, last modified January 29, 2022, en.wikipedia.org/wiki/Wright Flyer. See also "Boeing 747," Wikipedia, last modified January 29, 2022, en.wikipedia.org/wiki/Boeing_747.

5 "NOVA | Wright Brothers' Flying Machine | Pilot the 1903 Flyer (Non-Interactive)," PBS, September 2003, www.pbs.org/wgbh/nova/wright/flye-nf.html.

6 "Boeing 737 Facts," Boeing Commercial Airplanes, last modified April 2014, www.boeing.com/farnborough2014/pdf/BCA/fct%20-737%20Family%20Facts.pdf.

7 Daniel Keyes, "Facial recognition payments could overtake QR codes in China," Business Insider, November 20, 2019, www.businessinsider.com/facial-recognition-payments-could-break-out-in-china-2019-11.

8 Dave Makichuk, "PLA using 'exoskeleton suits' on Himalayan border," Asia Times, December 12, 2020, asiatimes.com/2020/12/pla-takes-high-ground-with-exoskeleton-suits-report/.

9 James Chen, "Bretton Woods Agreement and System," Investopedia, last modified April 28, 2021, www.investopedia.com/terms/b/brettonwoodsagreement.asp. See also Richard Best, "How the U.S. Dollar Became the World's Reserve Currency," Investopedia, last modified September 22, 2021, www.investopedia.com/articles/forex-currencies/092316/how-us-dollar-became-worlds-reserve-currency.asp.

10 Paul Mozur, "Google's AlphaGo Defeats Chinese Go Master in Win for A.I.," The New York Times, May 23 2017, www.nytimes.com/2017/05/23/business/google-deepmind-alphago-go-champion-defeat.html.

11 Cade Metz, The Genius Makers: The Mavericks Who Brought A.I. to Google, Facebook, and the World (New York: Random House, 2021, pp. 222).

12 Larry Hardesty, "Explained: Neural networks," MIT News, April 14, 2017, news.mit.edu/2017/explained-neural-networks-deep-learning-0414.

13 Cade Metz, The Genius Makers: The Mavericks Who Brought A.I. to Google, Facebook, and the World (New York: Random House, 2021, pp. 30-31).

14 David Gilbert, "Google's new products are proof AI computing has finally come of age," Yahoo! News, October 5, 2016, https://ca.news.yahoo.com/googles-products-proof-ai-computing-171957069.html.

15 "Google Strategy Teardown: Google Is Turning Itself into An AI Company as It Seeks to Win New Markets Like Cloud and Transportation," CB Insights Research, July 7, 2020, www.cbinsights.com/research/report/google-strategy-teardown/.

16 Matt Burgess, "Now DeepMind's AI can spot eye disease just as well as your doctor," WIRED UK, August 13, 2018, https://www.wired.co.uk/article/deepmind-moorfields-ai-eye-nhs.
17 Harvard Business Review, Artificial Intelligence: The Insights You Need from Harvard Business Review (HBR Insights Series) (Brighton: Harvard Business Review Press, 2019, pp. 34-35).
18 Ben Dickson, "Why Microsoft's new AI acquisition is a big deal," VentureBeat, April 17, 2021, venturebeat.com/2021/04/17/why-microsofts-new-ai-acquisition-is-a-big-deal/.
19 "Company Overview," Baidu Inc, https://ir.baidu.com/company-overview.
20 Edward Tse, China's Disruptors: How Alibaba, Xiaomi, Tencent and Other Companies Are Changing the Rules of Business (London: Portfolio, 2015, Chapter 1).
21 Waiyee Yip, "Singles Day: The world's biggest shopping event luring scammers," BBC News, November 11, 2020, www.bbc.com/news/world-asia-china-54898680.
22 Edward Tse, China's Disruptors: How Alibaba, Xiaomi, Tencent and Other Companies Are Changing the Rules of Business (London: Portfolio, 2015, pp. 85-86).
23 Lauren deLisa Coleman, "Get Ready! Here's How Tech Behemoth Tencent Is About To Up The AI Ante," Forbes, June 23, 2019, www.forbes.com/sites/laurencoleman/2019/06/23/get-ready-heres-how-tech-behemoth-tencent-is-about-to-up-the-ai-ante/.
24 Bernard Marr, "Artificial Intelligence (AI) In China: The Amazing Ways Tencent Is Driving It's Adoption," Forbes, June 4, 2018, www.forbes.com/sites/bernardmarr/2018/06/04/artificial-intelligence-ai-in-china-the-amazing-ways-tencent-is-driving-its-adoption/.
25 Elsa B. Kania, "Battlefield Singularity: Artificial Intelligence, Military Revolution, and China's Future Military Power," JSTOR, November 1, 2017, http://www.jstor.com/stable/resrep16985.6
26 Christian Brose, The Kill Chain: Defending America in the Future of High-Tech Warfare, (Paris: Hachette Books, 2020, pp. 56-57).

27 "Army War College Strategy Conference." U.S. Department of Defense, https://www.defense.gov/News/Speeches/Speech/Article/606661/army-war-college-strategy-conference/).

28 Christian Brose, The Kill Chain: Defending America in the Future of High-Tech Warfare, (Paris: Hachette Books, 2020, pp. 42).

29 Christian Brose, The Kill Chain: Defending America in the Future of High-Tech Warfare, (Paris: Hachette Books, 2020, pp. 47).

30 Christian Brose, The Kill Chain: Defending America in the Future of High-Tech Warfare, (Paris: Hachette Books, 2020).

31 Elsa B. Kania, "Battlefield Singularity: Artificial Intelligence, Military Revolution, and China's Future Military Power," Center for a New American Society, November 28, 2017, https://www.cnas.org/publications/reports/battlefield-singularity-artificial-intelligence-military-revolution-and-chinas-future-military-power

32 Elsa B. Kania, "Battlefield Singularity: Artificial Intelligence, Military Revolution, and China's Future Military Power," Center for a New American Society, November 28, 2017, pp. 155

33 Elsa B. Kania, "Battlefield Singularity: Artificial Intelligence, Military Revolution, and China's Future Military Power," Center for a New American Society, November 28, 2017, pp. 157

34 Elsa B. Kania, "Battlefield Singularity: Artificial Intelligence, Military Revolution, and China's Future Military Power," Center for a New American Society, November 28, 2017, pp. 220

35 Elsa B. Kania, "Battlefield Singularity: Artificial Intelligence, Military Revolution, and China's Future Military Power," Center for a New American Society, November 28, 2017, pp.:221

36 Elsa B. Kania, "Battlefield Singularity: Artificial Intelligence, Military Revolution, and China's Future Military Power," Center for a New American Society, November 28, 2017, pp. 223

37 Christian Brose, The Kill Chain: Defending America in the Future of High-Tech Warfare, (Paris: Hachette Books, 2020 pp. 63-64).

38 Christian Brose, The Kill Chain: Defending America in the Future of High-Tech Warfare, (Paris: Hachette Books, 2020, pp. 58).

39 George Galdorisi, AI at War: How Big Data Artificial Intelligence and Machine Learning Are Changing Naval Warfare (Annapolis: Naval Institute Press, 2021, pp 83).

40 George Galdorisi, AI at War: How Big Data Artificial Intelligence and Machine Learning Are Changing Naval Warfare (Annapolis: Naval Institute Press, 2021, pp 85).

41 Christian Brose, The Kill Chain: Defending America in the Future of High-Tech Warfare, (Paris: Hachette Books, 2020 pp. 74).

42 Elsa B. Kania, "Battlefield Singularity: Artificial Intelligence, Military Revolution, and China's Future Military Power," Center for a New American Society, November 28, 2017, https://www.cnas.org/publications/reports/battlefield-singularity-artificial-intelligence-military-revolution-and-chinas-future-military-power.

43 Nigel Inkster, The Great Decoupling: China, America and the Struggle for Technological Supremacy (London: Hurst, 2021, pp. 210-211).

44 Eric J. Topol, Deep Medicine: How Artificial Intelligence Can Make Healthcare Human Again, (New York: Basic Books, 2019, pp. 114–115).

45 Eric J. Topol, Deep Medicine: How Artificial Intelligence Can Make Healthcare Human Again, (New York: Basic Books, 2019, pp. 114–115).

46 Eric J. Topol, Deep Medicine: How Artificial Intelligence Can Make Healthcare Human Again, (New York: Basic Books, 2019, pp. 117).

47 Eric J. Topol, Deep Medicine: How Artificial Intelligence Can Make Healthcare Human Again, (New York: Basic Books, 2019, pp. 181-182).

48 Sara Castellanos, "How AI Played a Role in Pfizer's COVID-19 Rollout," The Wall Street Journal, April 1, 2021, https://www.wsj.com/articles/how-ai-played-a-role-in-pfizers-covid-19-vaccine-rollout-11617313126.

49 Thomas Davenport, "The potential for artificial intelligence in healthcare," Future Healthcare Journal, June 6, 2019, https://www.ncbi.nlm.nih.gov/pmc/articles/PMC6616181/.

50 Kai-Fu Lee, "Tech companies should stop pretending AI won't destroy jobs," MIT Technology Review, February 21, 2018, https://www.technologyreview.com/2018/02/21/241219/tech-companies-should-stop-pretending-ai-wont-destroy-jobs/.

51 Gil Press, "The Future Of AI In Healthcare," Forbes, April 29, 2021, https://www.forbes.com/sites/gilpress/2021/04/29/the-future-of-ai-in-healthcare/?sh=7507a728163b.

52 Lingling Wei, "China's New Power Play: More Control of Tech Companies' Troves of Data," The Wall Street Journal, June 12, 2021, https://www.wsj.com/articles/chinas-new-power-play-more-control-of-tech-companies-troves-of-data-11623470478.

53 Tom Simonite, "How Health Care Data and Lax Rules Help China Prosper in AI," Wired, January 8, 2019, https://www.wired.com/story/health-care-data-lax-rules-help-china-prosper-ai/.

54 Neal E. Boudette and Coral Davenport, "G.M. Will Sell Only Zero-Emission Vehicles by 2035," The New York Times, last modified October 1, 2021, https://www.nytimes.com/2021/01/28/business/gm-zero-emission-vehicles.html.

55 Lawrence D. Burns and Christopher Shulgan, Autonomy: The Quest to Build a Driverless Car—And How It Will Reshape Our World, (New York: Ecco, 2018, pp. 2-9).

56 Jeffrey B. Greenblatt and Samveg Saxena, "Autonomous taxis could greatly reduce greenhouse gas emissions of US light-duty vehicles," Nature Climate Change 5, July 6, 2015, pp. 860-863.

57 José Viegas, Luis Martinez, and Philippe Crist, "Shared Mobility: Innovation for Livable Cities," International Transport Forum Corporate Partnership Board, May 9, 2016, https://www.itf-oecd.org/sites/default/files/docs/shared-mobility-liveable-cities.pdf

58 Lawrence D. Burns and Christopher Shulgan, Autonomy: The Quest to Build a Driverless Car—And How It Will Reshape Our World, (New York: Ecco, 2018, pp. 223-225).

59 Daniel Sperling, Three Revolutions: Steering Automated, Shared, and Electric Vehicles to a Better Future, (Washington: Island Press, 2018, pp. 28).

60 "Tesla Model S," 2013 Tesla Model S Reviews, Ratings, Prices—Consumer Reports, https://www.consumerreports.org/cars/tesla/model-s/2013/overview/.

61 Drew Desilver, "Today's electric vehicle market: Slow growth in U.S., faster in China, Europe," Pew Research Center, June 7, 2021, https://www.pewresearch.org/fact-tank/2021/06/07/todays-electric-vehicle-market-slow-growth-in-u-s-faster-in-china-europe/.

62 Michael Schuman, "The Electric-Car Lesson That China Is Serving Up for America," The Atlantic, May 21, 2021, https://www.theatlantic.com/international/archive/2021/05/joe-biden-china-infrastructure/618921/.

63 Daniel Sperling, Three Revolutions: Steering Automated, Shared, and Electric Vehicles to a Better Future, (Washington: Island Press, 2018, pp. 179).

64 John Zimmer, "The Third Transportation Revolution: Lyft's Vision for the Next Ten Years and Beyond," Medium, September 18, 2016, https://medium.com/@johnzimmer/the-third-transportation-revolution-27860f05fa91.

65 James Crabtree, "Didi Chuxing took on Uber and won. Now it's taking on the world," Wired, September 2, 2018, https://www.wired.co.uk/article/didi-chuxing-china-startups-uber.

66 James Crabtree, "Didi Chuxing took on Uber and won. Now it's taking on the world," Wired, September 2, 2018, https://www.wired.co.uk/article/didi-chuxing-china-startups-uber.

67 Catherine Shu, "Baidu announces Apollo Enterprise, its new platform for mass-produced autonomous vehicles," TechCrunch, January 9, 2019, https://techcrunch.com/2019/01/08/baidu-announces-apollo-enterprise-its-new-platform-for-mass-produced-autonomous-vehicles/?guccounter=1&guce_referrer=aHR0cHM6Ly93d3cuZ29vZ2xlLmNvbS8&guce_referrer_sig=AQAAALp01jhMrStMi-JARqes66QLDNLOCW2hNpQohPsejOJ0fRMriF3fY28no4IclA-jDutwwml5gtrTD6LFV63W3vZfjyaev.

68 Michael E. McGrath, Autonomous Vehicles: Opportunities, Strategies, and Disruptions, (Independently published, 2018, pp. 244-245).

69 Christian Kurzydlowski, "The VR Market in China: Moving Toward the Metaverse," The China Guys, August 12, 2021, https://thechina-guys.com/china-virtual-reality-market/.

70 Mark Crawford, "10 Best IoT Examples in 2020," The American Society of Mechanical Engineers, February 19, 2020, https://www.asme.org/topics-resources/content/10-best-iot-examples-in-2020.

71 Long, Tony. "Aug. 29, 1949: First Soviet Atomic Test Stuns West." Wired, Conde Nast, 5 June 2017, www.wired.com/2007/08/dayintech-0829/

72 Lawn, Victor H. "Satellite Rivals Sirius Over City." Archives - The New York Times, The New York Times, 8 Nov. 1957, _archive.nytimes.com/www.nytimes.com/partners/aol/special/sputnik/sput-19.html (Publishing date listed is original print publication date.)

73 Ethen K. Lieser, "Why Didn't the Soviets Ever Make It to the Moon?" The National Interest, April 16, 2020, nationalinterest.org/blog/techland/why-didn%E2%80%99t-soviets-ever-make-it-moon-144992.

74 Prestowitz, p. 22

75 "Living in China's Technological Miracle," Global Times, July 29, 2021, https://www.globaltimes.cn/page/202107/1230033.shtml

76 Steven Lee Myers and Kenneth Chang, "China Brings Moon Rocks to Earth, and a New Era of Competition to Space," The New York Times, December 16, 2020, www.nytimes.com/2020/12/16/science/china-moon-mission-rocks.html.

77 Katherine Stapleton, "China now produces twice as many graduates a year as the US," World Economic Forum, April 13, 2017, www.weforum.org/agenda/2017/04/higher-education-in-china-has-boomed-in-the-last-decade.